IMMOBILIZED BIOCHEMICALS
AND AFFINITY CHROMATOGRAPHY

ADVANCES IN EXPERIMENTAL MEDICINE AND BIOLOGY

Editorial Board:

Nathan Back	Chairman, Department of Biochemical Pharmacology, School of Pharmacy, State University of New York, Buffalo, New York
N. R. Di Luzio	Chairman, Department of Physiology, Tulane University School of Medicine, New Orleans, Louisiana
Bernard Halpern	Collège de France, Director of the Institute of Immuno-Biology, Paris, France
Ephraim Katchalski	Department of Biophysics, The Weizmann Institute of Science, Rehovoth, Israel
David Kritchevsky	Wistar Institute, Philadelphia, Pennsylvania
Abel Lajtha	New York State Research Institute for Neurochemistry and Drug Addiction, Ward's Island, New York
Rodolfo Paoletti	Institute of Pharmacology and Pharmacognosy, University of Milan, Milan, Italy

Volume 1
THE RETICULOENDOTHELIAL SYSTEM AND ATHEROSCLEROSIS
Edited by N. R. Di Luzio and R. Paoletti • 1967

Volume 2
PHARMACOLOGY OF HORMONAL POLYPEPTIDES AND PROTEINS
Edited by N. Back, L. Martini, and R. Paoletti • 1968

Volume 3
GERM-FREE BIOLOGY: Experimental and Clinical Aspects
Edited by E. A. Mirand and N. Back • 1969

Volume 4
DRUGS AFFECTING LIPID METABOLISM
Edited by W. L. Holmes, L. A. Carlson, and R. Paoletti • 1969

Volume 5
LYMPHATIC TISSUE AND GERMINAL CENTERS IN IMMUNE RESPONSE
Edited by L. Fiore-Donati and M. G. Hanna, Jr. • 1969

Volume 6
RED CELL METABOLISM AND FUNCTION
Edited by George J. Brewer • 1970

Volume 7
SURFACE CHEMISTRY OF BIOLOGICAL SYSTEMS
Edited by Martin Blank • 1970

Volume 8
BRADYKININ AND RELATED KININS: Cardiovascular, Biochemical, and Neural Actions
Edited by F. Sicuteri, M. Rocha e Silva, and N. Back • 1970

Volume 9
SHOCK: Biochemical, Pharmacological, and Clinical Aspects
Edited by A. Bertelli and N. Back • 1970

Volume 10
THE HUMAN TESTIS
Edited by E. Rosemberg and C. A. Paulsen • 1970

Volume 11
MUSCLE METABOLISM DURING EXERCISE
Edited by B. Pernow and B. Saltin • 1971

Volume 12
MORPHOLOGICAL AND FUNCTIONAL ASPECTS OF IMMUNITY
Edited by K. Lindahl-Kiessling, G. Alm, and M. G. Hanna, Jr. • 1971

Volume 13
CHEMISTRY AND BRAIN DEVELOPMENT
 Edited by R. Paoletti and A. N. Davison • 1971

Volume 14
MEMBRANE-BOUND ENZYMES
 Edited by G. Porcellati and F. di Jeso • 1971

Volume 15
THE RETICULOENDOTHELIAL SYSTEM AND IMMUNE PHENOMENA
 Edited by N. R. Di Luzio and K. Flemming • 1971

Volume 16A
THE ARTERY AND THE PROCESS OF ARTERIOSCLEROSIS: Pathogenesis
 Edited by Stewart Wolf • 1971

Volume 16B
THE ARTERY AND THE PROCESS OF ARTERIOSCLEROSIS: Measurement and Modification
 Edited by Stewart Wolf • 1971

Volume 17
CONTROL OF RENIN SECRETION
 Edited by Tatiana A. Assaykeen • 1972

Volume 18
THE DYNAMICS OF MERISTEM CELL POPULATIONS
 Edited by Morton W. Miller and Charles C. Kuehnert • 1972

Volume 19
SPHINGOLIPIDS, SPHINGOLIPIDOSES AND ALLIED DISORDERS
 Edited by Bruno W. Volk and Stanley M. Aronson • 1972

Volume 20
DRUG ABUSE: Nonmedical Use of Dependence-Producing Drugs
 Edited by Simon Btesh • 1972

Volume 21
VASOPEPTIDES: Chemistry, Pharmacology, and Pathophysiology
 Edited by N. Back and F. Sicuteri • 1972

Volume 22
COMPARATIVE PATHOPHYSIOLOGY OF CIRCULATORY DISTURBANCES
 Edited by Colin M. Bloor • 1972

Volume 23
THE FUNDAMENTAL MECHANISMS OF SHOCK
 Edited by Lerner B. Hinshaw and Barbara G. Cox • 1972

Volume 24
THE VISUAL SYSTEM: Neurophysiology, Biophysics, and Their Clinical Applications
 Edited by G. B. Arden • 1972

Volume 25
GLYCOLIPIDS, GLYCOPROTEINS, AND MUCOPOLYSACCHARIDES OF THE NERVOUS SYSTEM
 Edited by Vittorio Zambotti, Guido Tettamanti, and Mariagrazia Arrigoni • 1972

Volume 26
PHARMACOLOGICAL CONTROL OF LIPID METABOLISM
 Edited by William L. Holmes, Rodolfo Paoletti, and David Kritchevsky • 1972

Volume 27
DRUGS AND FETAL DEVELOPMENT
 Edited by M. A. Klingberg, A. Abramovici, and J. Chemke • 1973

Volume 28
HEMOGLOBIN AND RED CELL STRUCTURE AND FUNCTION
 Edited by George J. Brewer • 1972

Volume 29
MICROENVIRONMENTAL ASPECTS OF IMMUNITY
 Edited by Branislav D. Jankovic and Katarina Isakovic • 1972

Volume 30
HUMAN DEVELOPMENT AND THE THYROID GLAND: Relation to Endemic Cretinism
   Edited by J. B. Stanbury and R. L. Kroc • 1972

Volume 31
IMMUNITY IN VIRAL AND RICKETTSIAL DISEASES
   Edited by A. Kohn and M. A. Klingberg • 1973

Volume 32
FUNCTIONAL AND STRUCTURAL PROTEINS OF THE NERVOUS SYSTEM
   Edited by A. N. Davison, P. Mandel, and I. G. Morgan • 1972

Volume 33
NEUROHUMORAL AND METABOLIC ASPECTS OF INJURY
   Edited by A. G. B. Kovach, H. B. Stoner, and J. J. Spitzer • 1972

Volume 34
PLATELET FUNCTION AND THROMBOSIS: A Review of Methods
   Edited by P. M. Mannucci and S. Gorini • 1972

Volume 35
ALCOHOL INTOXICATION AND WITHDRAWAL: Experimental Studies
   Edited by Milton M. Gross • 1973

Volume 36
RECEPTORS FOR REPRODUCTIVE HORMONES
   Edited by Bert W. O'Malley and Anthony R. Means • 1973

Volume 37A
OXYGEN TRANSPORT TO TISSUE: Instrumentation, Methods, and Physiology
   Edited by Haim I. Bicher and Duane F. Bruley • 1973

Volume 37B
OXYGEN TRANSPORT TO TISSUE: Pharmacology, Mathematical Studies, and Neonatology
   Edited by Haim I. Bicher and Duane F. Bruley • 1973

Volume 38
HUMAN HYPERLIPOPROTEINEMIAS: Principles and Methods
   Edited by R. Fumagalli, G. Ricci, and S. Gorini • 1973

Volume 39
CURRENT TOPICS IN CORONARY RESEARCH
   Edited by Colin M. Bloor and Ray A. Olsson • 1973

Volume 40
METAL IONS IN BIOLOGICAL SYSTEMS: Studies of Some Biochemical and Environmental Problems
   Edited by Sanat K. Dahr • 1973

Volume 41A
PURINE METABOLISM IN MAN: Enzymes and Metabolic Pathways
   Edited by O. Sperling, A. De Vries, and J. B. Wyngaarden • 1974

Volume 41B
PURINE METABOLISM IN MAN: Biochemistry and Pharmacology of Uric Acid Metabolism
   Edited by O. Sperling, A. De Vries, and J. B. Wyngaarden • 1974

Volume 42
IMMOBILIZED BIOCHEMICALS AND AFFINITY CHROMATOGRAPHY
   Edited by R. B. Dunlap • 1974

Volume 43
ARTERIAL MESENCHYME AND ARTERIOSCLEROSIS
   Edited by William D. Wagner and Thomas B. Clarkson • 1974

Volume 44
CONTROL OF GENE EXPRESSION
   Edited by Alexander Kohn and Adam Shatkay • 1974

Volume 45
THE IMMUNOGLOBULIN A SYSTEM
   Edited by Jiri Mestecky and Alexander R. Lawton • 1974

IMMOBILIZED BIOCHEMICALS AND AFFINITY CHROMATOGRAPHY

Edited by
R. Bruce Dunlap
Department of Chemistry
University of South Carolina
Columbia, South Carolina

PLENUM PRESS • NEW YORK AND LONDON

Library of Congress Cataloging in Publication Data

Symposium on Affinity Chromatography and Immobilized Biochemicals, Charleston, S. C., 1973.
Immobilized biochemicals and affinity chromatography.

(Advances in experimental medicine and biology, v. 42)
Contains most of the papers presented at the symposium held Nov. 7-9 in conjunction with the Southeastern Regional American Chemical Society meeting.
Includes bibliographical references.
1. Biological chemistry — Technique — Congresses. 2. Affinity chromatography — Congresses. I. Dunlap, Robert Bruce, 1942- ed. II. Series. [DNLM: 1. Biochemistry — Congresses. 2. Chromatography, Affinity — Congresses W1 AD559 v. 42 1974 / QD271 S986i 1973.]
QP519.7.S9   1973   574.1'92'028   74-7471
ISBN 0-306-39042-6

Proceedings of the symposium on Affinity Chromatography and Immobilized Biochemicals held in Charleston, South Carolina, November 7 - 9, 1973

© 1974 Plenum Press, New York
A Division of Plenum Publishing Corporation
227 West 17th Street, New York, N.Y. 10011

United Kingdom edition published by Plenum Press, London
A Division of Plenum Publishing Company, Ltd.
4a Lower John Street, London W1R 3PD, England

All rights reserved

No part of this book may be reproduced, stored in a retrieval system, or transmitted, in any form or by any means, electronic, mechanical, photocopying, microfilming, recording, or otherwise, without written permission from the Publisher

Printed in the United States of America

Preface

This volume contains most of the papers presented at the Symposium on Affinity Chromatography and Immobilized Biochemicals, which was held on November 7-9, 1973 in Charleston, South Carolina in conjunction with the Southeastern Regional American Chemical Society meeting. The topics of the symposium represent two new biochemical frontiers which have emerged in recent years through the ingenious development and application of solid phase biochemical technologies. Affinity chromatography involves the use of selected ligands, covalently bound to a solid support such as cellulose, glass, Sepharose, or polyacrylamide and exploits the principle of biochemical recognition between the ligand in the solid phase and a selected macromolecule to facilitate the rapid and often quantitative purification of enzymes, antibodies, antigens, hormones, other proteins, etc. The area of immobilized biochemicals includes the use of coenzymes, oligo- and polynucleotides, enzymes, and multistep enzyme systems which are immobilized or entrapped in the solid phase. The goals of the symposium were the review of the status of affinity chromatography and immobilized biochemicals, the presentation of new data and ideas in both areas, and the establishment of a dialogue between research workers in these two evolving disciplines which are so closely interrelated. The papers published in this volume provide the reader with reviews of several topics inherent in the solid phase biochemistry together with a series of timely manuscripts concerning new techniques and applications both in the use of affinity chromatography and in the investigation of immobilized biochemicals.

The symposium was international in flavor and attracted well over two hundred attendees. Support for the symposium was kindly and generously provided by Beckman Instruments, Bio-Rad Laboratories, Miles Laboratories-Research Division, Pharmacia Fine Chemicals, the Worthington Biochemical Corporation, the American Chemical Society, and the National Science Foundation.

I would like to express my thanks to the financial supporters, the presiding officers, and the participants who all contributed to the success of the symposium. Special thanks are due to Professor O. D. Bonner for his assistance in planning the symposium and to Miss Susan Carroll and Mrs. Anne Davis for their secretarial help performed during the planning phases of the symposium as well as in preparing manuscripts for the publisher.

January, 1974   R. Bruce Dunlap

Contents

PART ONE: AFFINITY CHROMATOGRAPHY

Affinity Chromatography--Old Problems and
New Approaches 3
      Steven C. March, Indu Parikh,
      and Pedro Cuatrecasas

Affinity Chromatography. New Approaches for the
Preparation of Spacer Containing Derivatives and for
Specific Isolation of Peptides 15
      Meir Wilchek

Quantitative Parameters in Affinity Chromatography 33
      A. H. Nishikawa, P. Bailon, and A. H. Ramel

Non-Specific Binding of Proteins by Substituted
Agaroses 43
      B. H. J. Hofstee

A Solid Phase Radioimmune Assay for Ornithine
Transcarbamylase 61
      Donald L. Eshenbaugh, Donald Sens
      and Eric James

Purification of Acetylcholinesterase by
Covalent Affinity Chromatography 75
      Houston F. Voss, Y. Ashani, and
      Irwin B. Wilson

Cooperative Effects of AMP, ATP, and Fructose
1,6-Diphosphate on the Specific Elution of Fructose
1,6-Diphosphatase from Cellulose Phosphate 85
      Joseph Mendicino and Hussein Abou-Issa

An Analysis of Affinity Chromatography Using
Immobilised Alkyl Nucleotides 99
      P. D. G. Dean, D. B. Craven, M. J. Harvey,
      and C. R. Lowe

Affinity Chromatography of Kinases and Dehydrogenases on Sephadex and Sepharose Dye Derivatives 123
    Richard L. Easterday and Inger M. Easterday

Affinity Chromatography of Thymidylate Synthetases Using 5-Fluoro-2'-Deoxyuridine 5'-Phosphate Derivatives of Sepharose 135
    John M. Whitely, Ivanka Jerkunica, and Thomas Deits

The Biosynthesis of Riboflavin: Affinity Chromatography Purification of GTP-Ring-Opening Enzyme 147
    L. Preston Mercer and Charles M. Baugh

Purification of Tyrosine-Sensitive 3-Deoxy-D-Arabino-heptulosonate-7-Phosphate and Tyrosyl-tRNA Synthetase on Agarose Carrying Carboxyl-Linked Tyrosine 157
    Andrew R. Gallopo, Philip S. Kotsiopoulos, and Scott C. Mohr

Structural Requirement of Ligands for Affinity Chromatography Absorbents: Purification of Aldehyde and Xanthine Oxidases 165
    Albert E. Chu and Sterling Chaykin

PART TWO: IMMOBILIZED BIOCHEMICALS

Immobilized Polynucleotides and Nucleic Acids 173
    P. T. Gilham

Immobilized Cofactors and Multi-Step Enzyme-Systems 187
    Klaus Mosbach

Preparation, Characterization, and Applications of Enzymes Immobilized on Inorganic Supports 191
    H. H. Weetall

Lactase Immobilized on Stainless Steel and Other Dense Metal and Metal Oxide Supports 213
    M. Charles, R. W. Coughlin, B. R. Allen, E. K. Paruchuri, and F. X. Hasselberger

The Use of Membrane-Bound Enzymes in an Immobilized Enzyme Reactor 235
    Charles C. Worthington

CONTENTS

The Optimization of Porous Materials for Immobilized
Enzyme Systems 241
    David L. Eaton

Water Encapsulated Enzymes in an Oil-Continuous
Reactor: Kinetics and Reactivity 259
    R. I. Leavitt, F. X. Ryan, and W. P. Burgess

Analysis of Reactions Catalyzed by Polysaccharide-
Enzyme Derivatives in Packed Beds 269
    M. H. Keyes and F. E. Semersky

The Preparation of Microenvironments for Bound
Enzymes by Solid Phase Peptide Synthesis 283
    James B. Taylor and Harold E. Swaisgood

Optimization of Activities of Immobilized Lysozyme,
α-Chymotrypsin, and Lipase 293
    Rathin Datta and David F. Ollis

Chemical Modification of Mushroom Tyrosinase
for Stabilization to Reaction Inactivation 317
    David Letts and Theodore Chase, Jr.

Chain Refolding and Subunit Interactions in
Enzyme Molecules Covalently Bound to a Solid Matrix 329
    H. Robert Horton and Harold E. Swaisgood

Immobilization of Lipase to Cyanogen Bromide
Activated Polysaccharide Carriers 339
    Paul Melius and Bi-Chong Wang

Use of Immobilized Enzymes for Synthetic Purposes 345
    David L. Marshall

Index 369

PART I
AFFINITY CHROMATOGRAPHY

AFFINITY CHROMATOGRAPHY - OLD PROBLEMS AND NEW APPROACHES

Steven C. March, Indu Parikh, and Pedro Cuatrecasas

Johns Hopkins University School of Medicine

Baltimore, Maryland 21205

The purification of macromolecules by biospecific adsorption-affinity chromatography (1) is a technique with a long history although its application to a wide variety of problems is comparatively recent. As long ago as 1910 there were reports of the purification of amylase by adsorption to insoluble starch (2). Tyrosinase was purified in 1953 using a cellulose matrix bearing the inhibitor, diazodizine (3). In both of these examples the power of purification lies in the highly selective "insolubilization" of the macromolecule from solution. This reversible phase separation often allows a high degree of concentration of the particular molecular species of interest from dilute solution and the removal, by washing and elution processes, of other molecular species that are cosolutes.

Affinity chromatography, like all chromatographic procedures, represents the application of the law of mass action. Enzyme kinetics can also be expressed, assuming pseudo-equilibrium, in terms of this law (Figure 1). Usually substrates are not used as affinity ligands (1a) although they may prove useful under conditions where catalysis proceeds at a very slow rate while binding is still high, e.g. at pH values distant from the pH optimum, or low temperatures. This hypothetical reaction sequence (1a) also includes a covalent intermediate and such intermediates have also proved useful in purifications (4). More frequently the coupled ligand used is an inhibitor (1b), a hormone (1c), and antibody (1d), or even an enzyme subunit (1e) as in studies on enzyme reconstitution. In these instances the physical parameter of most importance is the association (dissociation) constant. The magnitude of this quantity determines both whether the system will have a strong enough association to be useful and how one will have to treat the adsorbed material to recover it from the column.

a) $E + S \rightleftarrows E \cdot S \rightarrow E\text{-}S \rightarrow E + P$ E enzyme
   S substrate
b) $I + E \rightleftarrows I \cdot E$ I inhibitor
   H hormone
c) $H + R \rightleftarrows H \cdot R$ A antibody

d) $H + A \rightleftarrows H\text{-}A$

e) $nE \rightleftarrows E_n$

· indicates reversible complex
- indicates covalent intermediate

Figure 1

The transition, historically, from early isolated affinity purifications to the widespread usage we see now was made possible by three events. First was the publication, by Axen et al. (5), of a simple chemical procedure (CNBr) for activating polysaccharide matrices. Second was the commercial availability of a near ideal polysaccharide matrix, agarose, which possessed the necessary features of chemical and biological inertness, good chromatographic properties and ease of activation by cyanogen bromide (CNBr). Third, was the publication of several papers demonstrating that systematic approaches could be applied successfully to practical problems (6,7,8). After demonstration of the ease with which purifications could be accomplished using affinity systems, especially as compared to classical methods, and after a rational foundation was established for the design of absorbents, there was a very rapid increase in the use of affinity chromatography as a biochemical tool.

Enzyme purification continues to be the primary application of affinity chromatography. In this chapter, however, we would like to survey some of the more novel applications and to suggest some potential uses that have not yet appeared. During the purification of E. coli β-galactosidase it was noted that the catalytically inactive monomeric subunit was purified, by affinity chromatography, along with the active tetramer (6). This purification by binding, even when catalytic activity is lost through mutation or chemical modification, is an aid to studying catalytic mechanisms and structure-function relationships.

Particulate systems can also be analyzed by affinity chromatography. Hormones bound to agarose beads interact with fat cells and can stimulate biological activities similar to those observed following administration of the hormone in solution. Micrographs of fat cells sticking to agarose-insulin beads show that the interaction between the cells and beads includes changes in the cell shape resulting in maximization of the cell surface in contact with the bead (9). This active cell participation in the cell-bead interaction can result in such tight binding that the buoyant density of the cell-bead complex changes to the point that cells that would normally float to the top of the solution when centrifuged now sediment with the agarose (10).

Studies on lymphocytes predate the term affinity chromatography (11,12). For instance, glass beads coated with antigen can remove from a mixed lymphocyte sample those cells bearing the specific antibody directed against the coated antigen. Sensitized cells can be adsorbed to the beads and removed (13). Lymphocytes bearing anti-DNP, guinea pig albumin receptors were shown to be different from those cells bearing anti-DNP keyhole-limpet hemocyanin receptors indicating that different cell clones exist (14). More recently, lymphocytes sensitized to DNP antigens were removed from mixed cultures by binding to nylon fibers which had the DNP-antigen coupled to them. The fiber can be transferred to fresh medium and plucked to remove the bound cells (15).

Organelles can also be purified on affinity columns. Utilizing the ability of a newly synthesized enzyme to bind to its substrates even before the enzyme is released from the ribosome, polyribosomes capable of synthesizing tyrosine aminotransferase (TAT) were bound to a column containing coupled pyridoxal phosphate, a cosubstrate of the transfer reaction. A 100-fold purification of TAT synthesizing ribosomes was possible using this technique (16).

Viruses can also be purified (17). Using a neuraminidase inhibitor column a supposedly homogeneous preparation of influenza virus particles was purified two-fold through the interaction of a viral neuraminidase coat protein with the coupled ligand (18).

An exciting new area is the use of plant lectins as coupled ligands. Plant lectins are useful for several reasons. They allow purification on the basis of often complex carbohydrate structure and frequently simple sugars are effective competitors for binding, thus allowing easy elution of the bound material. Lectins have been used to purify blood group substances and glycoproteins like FSH and LH. Work has started on cell typing with coupled lectins and they are being used in studies on the differences between normal and neoplastic cells.

Affinity techniques have been applied to the purification of membrane constituents, virtually always in the presence of detergents. A solubilized fat cell membrane fraction was enriched for the insulin receptor using wheat germ agglutinin and Concanavalin A. Rhodopsin from bovine retinal cell outer rod membrane, can be purified on Concanavalin A columns (20). Glycopeptide purification and sequencing make use of a wide variety of lectins. Lectins may also prove useful in the separation of cytoplasmic membranes from intracellular membranes if their carbohydrate composition proves to be different.

Membrane receptors can also be purified using more functional ligands. The insulin receptor has been purified on insulin columns (21). Recently the cytoplasmic estradiol receptor has been purified on 17-substituted estradiol columns. These adsorbents were extremely effective in extracting the estradiol receptor only when very long spacer arms, e.g., albumin and poly-L-lysine-alanine copolymers were interposed between the sterol and the agarose matrix. Use of macromolecular spacer arms resulted in purification approaching 100,000-fold in one step with 30-50% recovery (22).

Returning again to soluble systems a vitamin B-12 binding protein has been isolated from human tissues on columns coupled to the partially hydrolyzed vitamin through amide bonds (23). The use of affinity techniques in radioimmunoassay is undergoing very rapid growth.

Another technique is the use of coupled ligands as probes of cell surfaces. When isoproterenol bearing beads are placed in contact with muscle cells there is a prompt inotropic effect which does not diminish with time as it would if the response were caused by the drug in solution (24). Control studies eliminated the possibility that ligand released from the beads was the cause of the inotropic effect. Removal of the beads resulted in prompt relaxation of the muscle.

Affinity chromatography is unique as a separation tool in that its application requires a good deal of knowledge of the biochemical system in which it is to be used before the affinity absorbent can be designed. The problem of selecting the matrix and ligand reoccurs with every new system approached. A useful ligand must have a high affinity (K_D typically 10^{-5} or greater) for the molecule that is to be purified. Having such a ligand, a point of chemical modification of the ligand must be available. The aim is to retain the biological binding activity of the ligand after it is coupled. Frequently one can study the effect of chemical modification of the potential ligand with soluble model systems. One can synthesize a soluble "armed" derivative, for example, and test its potency. Tests of this type may show that the modified ligand has an enhanced affinity supposedly due to hydrophobic contributions to binding resulting from the presence of the spacer. Enhancement of binding is not a guaranteed result of such modifications, however.

Coupling of the ligand to the matrix may involve reactions that are not completely defined. The use of gluteraldehyde as a coupling agent is an example of this. Even when the coupling chemistry is reasonably well understood, as in the reactions of CNBr activated agarose, the actual product may not be known. Protein ligands are assumed to couple the CNBr activated agarose through ϵ-amino groups. However, other polar groups may also react. Model studies done in our laboratory demonstrate that N-acetylated cysteine can compete with alanine in the coupling reaction. The blocked cysteine inhibits alanine binding by 50%, when present in 4-10 fold molar excess. The importance of sulfhydryl groups in coupling reactions involving proteins is unknown.

No coupling procedure is universally applicable. Some coupling methods are incompatible with useful matrices. Some coupling agents inactivate a large fraction of the ligand during coupling. One alternative that may not be immediately obvious is the possibility of activating the ligand to react with the matrix. The advantages of using an active ligand is that one can easily control the degree of activation, and the unreacted material is soluble and can be removed by washing. This may reduce the number of by-products coupled to the gel and may produce a more ideal matrix.

The most convenient method of matrix activation is no activation at all. Adsorption, like that of protein on glass, polymerization trapping, e.g. enzymes trapped in acrylamide gels, or binding to hydrophobic columns (25), requires no chemical manipulation of the ligand and, in some cases, produces matrices of useful short-term stability. Usually, however, some form of chemical activation is required to covalently couple molecules to the matrix. The most frequently used coupling reaction is the activation of agarose with CNBr. The intermediate cyanate produced is very unstable and rapidly reacts with hydroxyl radicals to form the more stable imidocarbonate. The imidocarbonate can then react with amino groups to form a variety of products, the major one probably being an isourea (26,27). Isoureas are protonated and positively charged at physiological pH. From this basic reaction a whole series of alternate pathways radiate. Amino compounds of all types, including compounds with reactive radicals can be coupled to modify the matrix for additional coupling reactions (28). Another frequently used matrix is polyacrylamide, which is also available in beaded form (29). Acrylamide is activated by treatment with hydrazine followed by nitrous acid to form reactive acyl azides. Potentially, polyacrylamine has a higher coupling capacity than agarose. The low porosity of the useful acrylamide beads has limited the utility of this matrix. The restricted porosity can be useful, however, in the preparation of particulate adsorbents. The importance of porosity is emphasized elsewhere in this volume in discussions on the use of glass as a matrix.

Activation of the matrix may produce some undesirable changes in the matrix properties. Frequently the substituted matrix is more hydrophobic. Many ionic residues may be introduced into a normally non-ionic polymer. Neither of these effects, however, has seriously impaired the use of such matrices. It should be mentioned that increases in hydrophobicity may increase binding in some cases and virtually all biological environments possess some ion-exchange properties. The degree of activation may be critical when single bond attachment is necessary for retention of activity. Most low molecular weight ligands can be designed to attach only at one point. Polymeric ligands may possess several reactive sites. Chemical modification of some of the sites may destroy the activity of the ligand. This is especially likely if the ligand couples to the matrix at more than one site. To lessen the likelihood of such complications lesser degrees of activation of the matrix may be necessary. Alternatively, the pH or temperature at which the coupling reaction is done can be adjusted to favor coupling at a particular site.

Leakage, release of the coupled ligand into the solvent, may be a problem. Leakage is important in at least two kinds of systems. Industrial usage often requires the use of high flow rates and high temperatures which accentuate otherwise minor leakage effects. The chapters in this volume on glass-enzyme reactors illustrate this effect. Secondly, systems of very high affinity and low concentrations of the desired macromolecule can be severely affected by the presence of relatively small amounts of released ligand. Hormone-receptor systems are an example of this group. If one obtains two micrograms of estrogen receptor protein in a homogenized calf uterus one can estimate that the molar concentration of receptor will be very low. The leakage of a few nanomoles of the high affinity ligand from the column could functionally inactivate all of the receptor in solution and prevent its binding to the column. Prevention of this effect often requires the dilution, with unsubstituted matrix, of the affinity matrix until there remains only enough ligand in the column to remove the species of interest from solution.

A model study performed on ligands coupled to agarose with CNBr and stored at 4° for 30 days suggested that about 0.1% of the alanine coupled was released per day. Albumin, coupled and stored under the same conditions eluted at a rate approximately 1/5 as fast (30). These columns were washed with 0.1 M carbonate buffer, pH 10, 2 M urea and 0.1 M acetate buffer, pH 4, all prepared in 0.5 M NaCl, before storage to reduce the amount of adsorbed ligand. Treatment of the coupled ligand with sodium borohydride to reduce the isourea did not reduce the rate of leakage. Methods of reducing leakage are discussed below.

The presence of a polyvalent matrix, such as is formed for affinity chromatography, may have pronounced effect on biological systems, even in the absence of any chemical effects. Incubation of insulin coupled to agarose beads with murine mammary cells rapidly causes increased transport of a non-metabolizable amino acid into the cells. Incubation of the same type of cells with insulin free in solution does not cause increased transport unless the cells have been preincubated for 24 hours with prolactin (31). This polyvalent affect may be related to the "patching" phenomenon observed with lymphocytes treated with fluorescent antibody. There may be a redistribution of surface receptors under the influence of the polymeric-coupled ligands. This effect may complicate the interpretation of experiments using affinity columns especially those involving particulate materials.

Once the ligand is coupled and the desired molecule is absorbed to the column, one has the final problem of elution. Elution by denaturation has several advantages. The reaction is rapid, leading to the prompt release of the protein in a small volume. The macromolecule must have the ability to renature for this method to be useful and this elution method is not specific. It only works well when there is only one or a few proteins retained by the column. Another common method of elution is by ligand competition. This requires the addition of a soluble ligand which competes for binding with the coupled ligand. If the affinity of the ligand-macromolecule complex is very high, dissociation kinetics may become important. For example, for high affinity complexes the dissociation rates are slow. One must allow five half-lives to pass before the "insoluble" complex is 95% dissociated. Elution may require incubation of the bound material with the competitive ligand for hours with zero flow conditions before dissociation is complete. Ligand competition elution provides the possibility for added specificity in that, by appropriate choice of competitive ligand, one of several adsorbed species may be selectively eluted. This method has been used in the resolution of dehydrogenases coupled to an NAD column (32). The bound molecules may also be removed by extruding the matrix from the column and treating the matrix in suspension. This dilutes the gel and pulls the equilibrium towards the dissociated species. Occassionally the macromolecule can be eluted without the addition of soluble ligand but in any case the kinetic considerations outlined above still are important. This technique results in very dilute solutions of the isolated molecule. Occassionally elution may require chemical or enzymic treatment of the bound-ligand macromolecule complex. An extreme example is the use of dextranase to digest Sephadex columns.

A new approach to the attachment of ligands to matrices is the use of polymeric arms or spacers. It is widely recognized that displacement of the ligand from the matrix by interposing a spacer-arm may markedly improve the effectiveness of the affinity ligand. Symmetric diamines and diacids have been very useful as spacer arms. Several short arms, a diamine, a diacid, and an amino-acid may for example be added sequentially to build up a longer multimeric spacer arm. Such multimeric arms were used in the purification of the estradiol receptor (8). Multimeric arms can be produced that bear an active functional group on the end distal to matrix. This group may react directly with amino-ligands. Active esters like N-hydroxy-succinimide can be synthesized, stored dry and added directly to solutions of the molecule to be coupled (33). This material is now commercially available (34). Another "activated" matrix, CNBr activated Sepharose, is also commercially available (35).

Recently macromolecular spacer arms have come into use. An example is polylysine-alanine. This material possesses a polylysine backbone whose ϵ-amino groups are substituted with 14 to 16 residue polyalanine peptides. This material has a molecular weight of about 260,000 and a Stokes radius of approximately 7.5 nm. This distance is important because, if one calculates the distance between attachment points in a polymeric matrix substituted to the extent of 10 nmoles per ml, the distance between attachment points is, on the average, 6 nm. The possibility exists therefore that polymeric arms like polylysine-alanine, can attach at two or more points to the matrix. All of the unattached amino terminals are available for the coupling of ligands using, for example, carbodiimides. This polymeric spacer-arm may be synthesized from D-amino acids. The D-polymer should be resistant to attack by most proteases.

Albumin, if coupled in the denatured state, can also be used as a macromolecular spacer arm. Albumin can be an "active" arm if the glutamate and aspartate residues are first converted to acyl-hydrazides. Treatment of hydrazido-albumin-agarose with nitrous acid allows coupling of amino-ligands (30). The potential importance of multipoint attachment of polymeric spacer arms is illustrated by considering the effect of such attachment on the rate of leakage. If a ligand leaks at the rate of 0.1% per day when coupled at one point and the cleavage of one attachment bond does not influence the rate of cleavage of a second attachment bond we may expect that a polymeric spacer arm attached at two points would leak at a rate of 0.0001% per day. The realization of this calculated effect can be seen in the utility of polymeric arms in the purification of the estradiol receptor where a multimeric arm allows a purification of 27-fold, use of albumin increases the purification 4,400-fold and a polylysine-alanine arm allows purification in excess of 100,000-fold (8).

Recently a modified method for activating agarose with CNBr

has been developed. Cyanogen bromide is dissolved in acetonitrile to a concentration of 2 gm per ml (19 M). Agarose is made up in a 50% v/v suspension in water. Equal volumes of the suspension and 2 M sodium carbonate are mixed and chilled to ice temperature. An appropriate volume of $CNBr/CH_3CN$ solution is added (0.2 gm CNBr per packed ml of agarose), the slurry rapidly stirred for 1-2 minutes, and the activated matrix washed and added to the ligand dissolved in 0.2 M sodium bicarbonate, pH 9.5 (36). This modified method reduces the hazards associated with weighing of CNBr and is more convenient than previously published techniques.

REFERENCES

1. Cuatrecasas, P. and Anfinsen, C. B. "Affinity Chromatography" An. Rev. Biochem. 40: 259, 1971.

2. Starkenstein, E. "Uber Fermentwirkung and deren Besiuflussung durch Neutralsalze" Biochem. Z. 24: 210, 1910.

3. Lerman, L. S. "A Biochemically Specific Method for Enzyme Isolation" Proc. Natl. Acad. Sci. USA 39: 232, 1953.

4. Ashani, Y. and Wilson, I. B. "A Covalent Affinity Column for the Purification of Acetylcholine-esterase" Biochim. Biophys. Acta 276: 317, 1972.

5. Axen, R. and Parath, J. "Chemical Coupling of Enzymes to Cross-Linked Dextran (Sephadex)" Nature 210: 367, 1966.

6. Cuatrecasas, P., Wilchek, M., Anfinsen, C. B. "Selective Enzyme Purification by Affinity Chromatography" Proc. Natl. Acad. Sci. USA 61: 636, 1968.

7. Steers, E., Cuatrecasas, P., and Pollard, H. "The Purification of β-Galactosidase from Escherichia coli by Affinity Chromatography" J. Biol. Chem. 246: 196, 1971.

8. Sica, V., Nola, E., Parikh, I., Puca, G. A., and Cuatrecasas, P. "Purification of Oestradiol Receptors by Affinity Chromatography" Nature New Biology 244: 36, 1973.

9. Cuatrecasas, P. "Interaction of Insulin with Cell Membrane-Primary Action of Insulin" Proc. Natl. Acad. Sci. USA 63: 450, 1969.

10. Soderman, D. D., Germershausen, J. and Katzen, H. M. "Affinity Binding of Intact Fat Cells and their Ghosts to Immobilized Insulin" Proc. Natl. Acad. Sci. USA 70: 792, 1973.

11. Henry, C., Kimura, J., and Wofsy, L. "Cell Separations on Affinity Columns - Isolation of Immunospecific Precursor Cells from Unimmunized Mice" Proc. Natl. Acad. Sci. USA 69: 34, 1972.

12. Truffa-Bachi, P. and Wofsy, L. "Specific Separation of Cells by Affinity Chromatography" Proc. Natl. Acad. Sci. USA 66: 685, 1970.

13. Singhal, S. K. and Wigzell, H. "In Vitro Induction of Specific Unresponsiveness of Immunologically Reactive Normal Bone Marrow Cells" J. Exptl. Med. 131: 149, 1970.

14. Davie, J. M. and Paul, W. F. "Receptors on Immunocompetent Cells. I. Receptor Specificity of Cells Participating in a Cellular Immune Response" Cell. Immunol. 1: 404, 1970.

15. Rutishauer, U., Millette, C. F. and Edelman, G. M. "Specific Fractionation of Immune Cell Populations" Proc. Natl. Acad. Sci. USA 69: 1596, 1972.

16. Miller, J. V., Cuatrecasas, P., and Thompson, E. B. "Partial Purification by Affinity Chromatography of Tyrosine Aminotransferase-Synthesizing Ribosomes from Hepatoma Tissue Culture Cells" Proc. Natl. Acad. Sci. USA 68: 1014, 1971.

17. Kenyon, A. J., Gander, J. E., Lopez, C., and Good, R. A. "Isolation of Aleutian Mink Disease Virus by Affinity Chromatography" Science 179: 187, 1973.

18. Cuatrecasas, P. and Illiano, G. "Purification of Neuraminidase from Vibrio Cholerea, Clostridium and Influenza Virus by Affinity Chromatography" Biochem. Biophys. Res. Comm. 44: 178, 1971.

19. Cuatrecasas, P. and Tell, G. P. E. "Insulin-Like Activity of Concanavalin A and Wheat-Germ Agglutinin - Direct Interactions with Insulin Receptors" Proc. Natl. Acad. Sci. USA 70: 485, 1973.

20. Steineman, A. and Stryer, L. "Accessibility of the Carbohydrate Moiety of Rhodopsin" Biochemistry 12: 1499, 1973.

21. Cuatrecasas, P. "Isolation of Insulin Receptor of Liver and Fat-Cell Membranes" Proc. Natl. Acad. Sci. USA 69: 316, 1972.

22. Sica, V., Parikh, I., Nola, F., Puca, G. A., and Cuatrecasas, P. "Affinity Chromatography and the Purification of Estrogen Receptors" J. Biol. Chem. 248: 6543, 1973.

23. Allen, R. H. and Majerus, P. W. "Isolation of Vitamin B_{12}-Binding Proteins Using Affinity Chromatography" J. Biol. Chem. 247: 7695, 1972.

24. Venter, J. C., Ross, J., Dixon, J. E., Mayer, S. E. and Kaplan, N. O. "Immobilized Catecholamines and Cocaine Effects on Contractility of Cardiac Muscle" Proc. Natl. Acad. Sci. USA 70: 1214, 1973.

25. Er-el, Z., Zaidenzaig, Y., and Shaltiel, S. "Hydrocarbon Coated Sepharoses. Use in the Purification of Glycogen Phosphorylase" Biochem. Biophys. Res. Comm. 49: 383, 1972.

26. Axen, R., Porath, M., and Ernback, S. "Chemical Coupling of Peptides and Proteins to Polysaccharides by Means of Cycrogen Halids" Nature 214: 1302, 1967.

27. Svensson, B. "Use of Isoelectric Focusing to Characterize the Bonds Established During Coupling of CNBr-Activated Amylodextrin to Subtilisin Type Novo" FEBS Letters 29: 167, 1973.

28. Cuatrecasas, P. "Protein Purification by Affinity Chromatography. Derivatizations of Agarose and Polyacrylamide Beads. J. Biol. Chem. 245: 3059, 1970.

29. Inman, J. K. and Dintzis, H. "The Derivatization of Cross-Linked Polyacrylamide Beads. Controlled Introduction of Functional Groups for the Preparation of Special-Purpose, Biochemical Adsorbents. Biochemistry 10: 4074, 1969.

30. Parikh, I., March, S. and Cuatrecasas, P. "Advances in the Methodology of Substitution Reactions and in the Use of Agarose in Affinity Chromatography" Methods in Enzymology 22 (B) Affinity Techniques (ed. W. Jakoby and M. Wilchek) Academic Press (in press).

31. Oka, T., Topper, Y. J. "Dynamics of Insulin Action on Mammary Epithelium" Nature New Biology 239: 216, 1972.

32. Lowe, C. R., Harvey, M. J., Craven, D. B., Kerfoot, M. A., Hollows, M. F., and Dean, P. D. G. "The Purification of Nicotinamide Nucleotide-Dependent Dehydrogenases on Immobilized Cofactors" Biochem. J. 133: 507, 1973.

33. Cuatrecasas, P. and Parikh, I. "Adsorbents for Affinity Chromatography. Use of N-Hydroxysuccinimide Esters of Agarose" Biochem. 11: 229, 1972.

34. BIO-RAD Laboratories, 32nd and Griffin Ave., Richmond, Ca. 94804.

35. Pharmacia Fine Chemicals Inc., 800 Centennial Ave., Piscataway, N. J. 08854.

36. March, S. C., Parikh, I., and Cuatrecasas, P. "A Simplified Method for Cyanogen Bromide Activation of Agarose for Affinity Chromatography" An. Biochem. (in press).

AFFINITY CHROMATOGRAPHY. NEW APPROACHES FOR THE PREPARATION OF

SPACER CONTAINING DERIVATIVES AND FOR SPECIFIC ISOLATION OF PEPTIDES

Meir Wilchek

Department of Biophysics
The Weizmann Institute of Science
Rehovot, Israel

STABLE, HIGH CAPACITY AND CHARGE-FREE AGAROSE DERIVATIVES

Affinity chromatography is a method for purification of biologically active compounds by virtue of biological recognition between certain ligands and proteins (1,2,3). Figure 1 shows a general scheme for purification of compounds by affinity chromatography. Purification is achieved by chromatography of a mixture containing the protein to be purified on a column of an insoluble matrix to which a specific ligand has been covalently bound. Proteins and other molecules not exhibiting appreciable affinity for the ligand will pass unretarded through the column, whereas those which "recognize" the ligand will be retarded. Elution of the bound protein, is readily achieved by changing the medium to conditions unfavourable for combination.

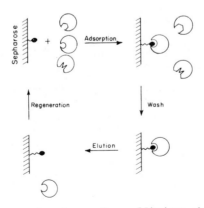

Fig. 1. A general scheme for affinity chromatography.

The successful application of the method requires that the essential group on the ligand for interaction with the molecules to be purified must be sufficiently distant from the matrix surface to relieve steric restrictions. This can be done by synthesizing a ligand with a long hydrocarbon chain and then attaching the elongated ligand to the agarose beads (1,4). Cuatrecasas (5) suggested that the use of aminoalkyl-Sepharose as a starting material. The amino derivative was then reacted with different functional groups, to which ligands were coupled.

All these derivatives have strong ion exchange properties due to incomplete substitution of the amino or carboxyl groups by the ligands. Fig. 2 shows an alternative approach for the preparation of agarose derivatives containing the same functional groups (6). These derivatives are free of ion exchange properties, resulting from coating the beads with diamines. This approach is preferable even though it is more complicated and requires some organic synthesis.

The nucleotide 3',5'-cyclic adenosine monophosphate (C-AMP) has been shown to exert many of its effects via activation of the enzyme protein kinases (7,8).

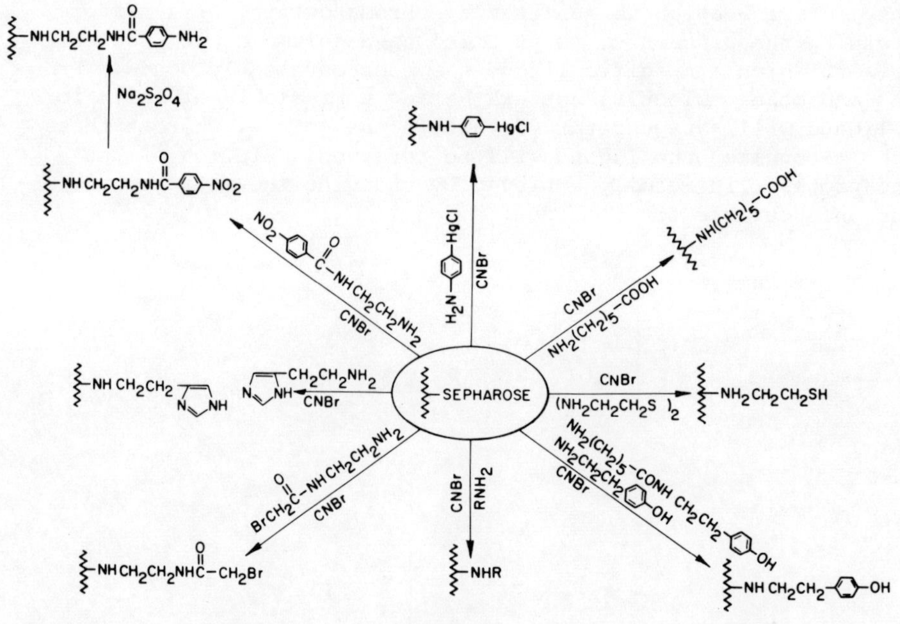

Fig. 2. Procedures for preparing derivatives of Sepharose.

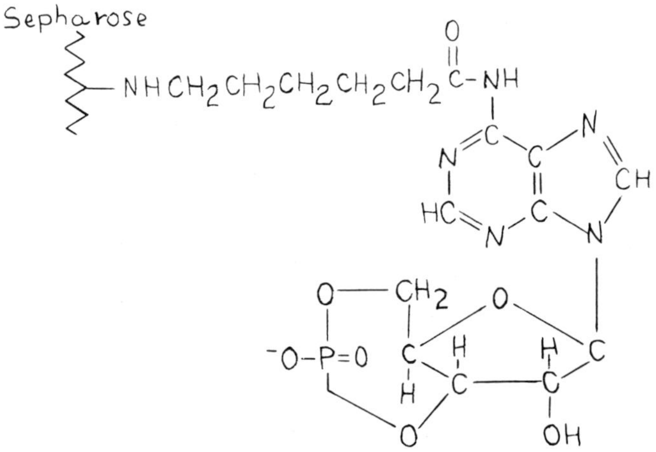

PREPARATION OF SPACER CONTAINING DERIVATIVES

Fig. 3

Recently, we have prepared a C-AMP-Sepharose column by coupling N^6-aminocaproyl 3',5'-cyclic adenosine monophosphate to cyanogen bromide activated Sepharose (Fig. 3). Chromatography of protein kinases on this column resulted in preparations which were fully active in the absence of C-AMP. C-AMP independent kinases thus no longer bind C-AMP (Table I). It was suggested that C-AMP Sepharose activated the protein kinase by removal of a regulatory unit which binds C-AMP and remained on the column.

TABLE I

CHROMATOGRAPHY OF PROTEIN KINASE ON CYCLIC AMP-SEPHAROSE COLUMNS

Fraction	Protein Kinase Activity ^{32}P Incorporated pmoles/mg protein/min		Recovery of Protein Kinase Activity	C-AMP Binding pmoles/mg Protein
	with C-AMP	without C-AMP	%	
I. Parotid supernatant	240	100	(100)	3.6
Column eluate	210	200	50	0.05
II. Muscle kinase	1470	70	(100)	12
Column eluate	2000	2000	70	0.1

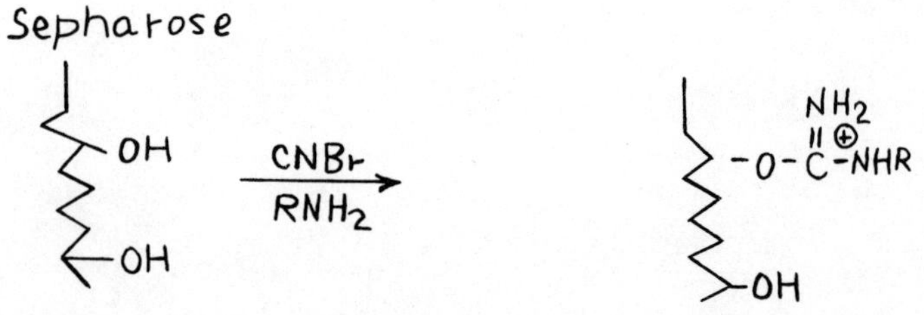

Fig. 4. N-substituted isourea formed on coupling of cyanogen bromide activated Sepharose with amines.

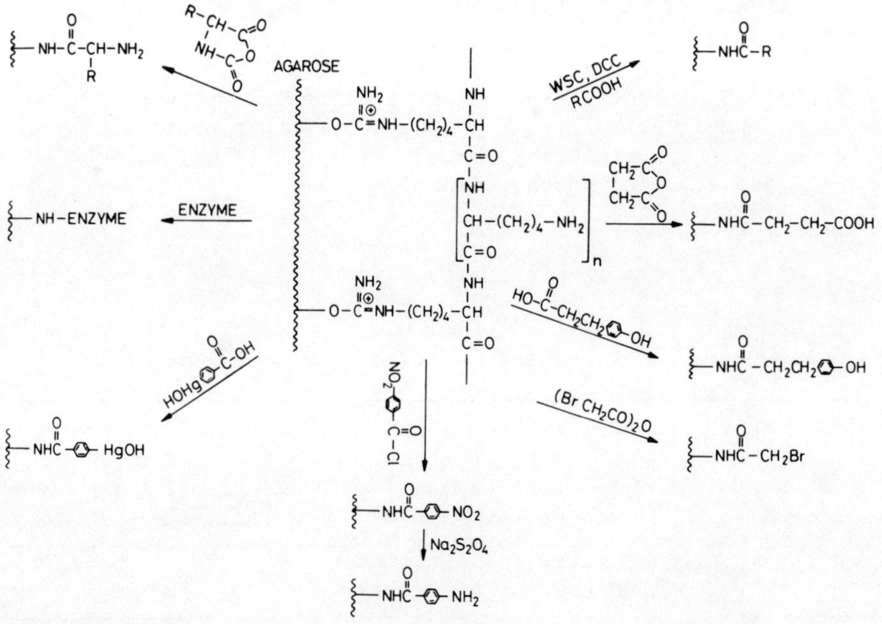

Fig. 5. Procedures for preparing derivatives of Sepharose-polylysine or Sepharose multi-poly-DL-alanine-polylysine.

This work was questioned by Tesser et al (10) since they prepared a different C-AMP Sepharose column and observed some leakage of the ligand from the column, therefore, they suggested that the effect observed by us may also be due to removal of enough cyclic AMP from the Sepharose together with the kinase and thus making it independent on C-AMP.

Such a leakage is not surprising since the bonds formed between cyanogen bromide activated Sepharose and amino groups are mainly N-substituted isoureas which are not completely stable (11,12)(Fig. 4). Therefore, these conjugates exhibit a small but constant leakage of the ligands from the solid matrix. This leakage could disallow the isolation of very small amounts of proteins by affinity chromatography, and would render the interaction between Sepharose bound hormones and intact cells questionable, especially peptide hormones bound through a single amino group.

The leakage of ligands from the column can be prevented by the use of a polyvalent spacer. Such a polyvalent spacer will render the agarose column highly stable due to the numerous positions at which the multivalent spacer will be coupled to the solid matrix. Stable and high capacity agarose derivatives were obtained by coupling polylysine, or multi-poly-DL-alanyl-polylysine to Sepharose (12, 14). The poly-DL-alanine was introduced to achieve structureless straight chains to serve as very long spacers between the solid matrix and the ligand. Another advantage of using poly-DL-alanine is that its amino group has a lower pK than the polylysine and can, therefore, be used more effectively in substitution reactions.

The polylysyl Sepharose and the poly-DL-alanyl-lysyl-Sepharose were used for further substitution with different functional groups in a variety of ways (Fig. 5). The methods of preparation of the various derivatives are essentially those described by Cuatrecasas (5), and will be published elsewhere.

The polylysyl-Sepharose conjugate is very stable, no leakage from polylysyl-Sepharose after reaction with fluorodinitrobenzene or dansyl chloride was observed even after storage for one year in a solution of sodium bicarbonate (pH 8) at room temperature.

Based on these observations we have prepared a stable new C-AMP-Sepharose column by coupling N^6-succinyl C-AMP to polylysine Sepharose. On passing protein kinases through this column the regulatory units were again adsorbed to the column (M.Wilchek, unpublished results).

All the agarose derivatives, including the monovalent and polyvalent-coupled derivatives have strong ion exchange properties, some due to incomplete coupling of the ligands to the amino or carboxyl groups. However, the main reason for the strong ion exchange properties of the agarose derivatives is the N-substituted isourea formed

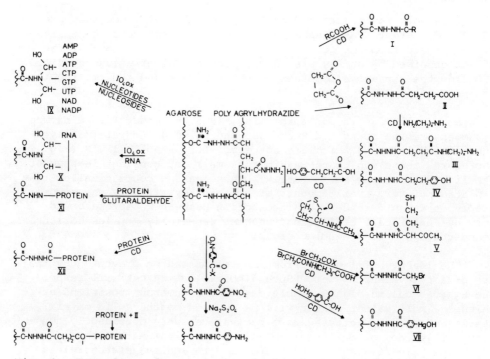

Fig. 6. Procedures for preparing derivatives of Sepharose-hydrazide

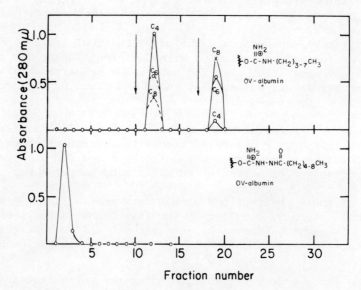

Fig. 7. Adsorption and elution patterns of ovalbumin on different alkyl-Sepharose and hydrazide Sepharoses. The adsorption was performed with 0.05 M Tris buffer.

on coupling amines to cyanogen bromide activated Sepharose. A pK of 10.4 was found by potentiometric titration of the isourea derivative formed on coupling butylamine to Sepharose. Thus the amino groups retain their basicity after reaction with Sepharose. These ion exchange effects will lower the degree of specific adsorption by the affinity column, and if a pH or salt gradient is used for elution the entire procedure is not more than simple ion exchange chromatography.

Furthermore, the hydrophobic nature of the spacer will also enhance the nonspecific adsorption capacity of the columns. Recently, it was shown that agarose derivatives containing the hydrocarbon spacer arms can adsorb some proteins (15,16,17). This kind of adsorption was termed "hydrophobic chromatography". The fact that binding of proteins occurred only at pH values higher than their isoelectric points, indicated that the adsorption resulted mostly from the positive charged N-substituted isourea, with some assistance by the hydrophobic side chains. This is also supported by the finding that the adsorbed proteins on the hydrophobic column can be eluted by high concentration of salt. It is generally accepted that hydrophobic interactions are strengthened rather than weakened by salts. Therefore, the adsorption of proteins by hydrocarbon coated agaroses should be attributed to reversible denaturation of the proteins by the "detergent like" agarose derivatives. This is suggested by the adsorption of penicillinase on an ethyl-Sepharose column, elution of the enzyme with 1 M sodium chloride gave an inactive enzyme (D. Shapiro and M.Wilchek, unpublished). In order to eliminate this interference to affinity chromatography, one should replace the hydrophobic spacer by an hydrophilic one, or prepare agarose derivatives free of charge. The second approach is preferable since hydrophobic interactions are usually weaker than electrostatic interaction.

Recently, we have shown that hydrazides can be coupled to cyanogen bromide activated agarose similar to the coupling of amines (18). The resulting derivatives are free of charge. In order to prepare agarose derivatives free of charge containing different functional groups for further ligand substitution, an excess of the following dihydrazides was reacted with Sepharose: $NH_2NHCO(CH_2)_nCO\ NHNH_2$. The reaction was performed either in bicarbonate or acetic acid solutions (19). These hydrazide agarose derivatives were used to prepare columns containing functional groups for further ligand substitutions (Figs. 2 and 6). Furthermore, to eliminate hydrophobic interactions completely, dihydrazides of tartaric acid, or malic acid can be used. But this is not necessary since we have found that proteins such as ovalbumin which are strongly adsorbed on butyl-Sepharose (C_4) are not adsorbed even on Sepharose derivatives to which decanoic acid hydrazide (10) was coupled (Fig. 7) (M.Wilchek and T.Miron).

The linkages between hydrazides and Sepharose are as stable as amines coupled monovalently to Sepharose and can be used safely for regular affinity chromatography. Thus NADP bound to a Sepharose hydrazide was used for complete adsorption of glucose-6-phosphate dehydrogenase (18,19) and ATP Sepharose hydrazide was used for adsorption and purification of heavy meromysin (19,20), without observing any leakage during operation.

If more stability is required polyhydrazide such as polyglutamic hydrazide (14) or linear polyacrylic hydrazide can be coupled to Sepharose. These high capacity agarose derivatives are very stable due to the multipoint attachment of the polyhydrazides to Sepharose.

The polyacrylic hydrazide-Sepharose derivatives, which are easy to prepare, behave as ideal insoluble carriers for affinity chromatography. They retain the properties of Sepharose; namely, minimal nonspecific interaction with proteins, good flow rate, and a very loose porous network which permits entry and exit of large molecules. In addition, they are mechanically and chemically stable to the conditions of coupling and elution. These derivatives, however, also retain the properties of polyacrylamide gels, in that they lack charged groups even after coupling the cyanogen bromide activated Sepharose (a pK of 4.2 was found on potentiometric titration of semicarbazide coupled to Sepharose) and they have a very large number of modifiable groups. These hydrazide-agarose derivatives, which can be considered "universal columns", can either be utilized directly (e.g. coupling of RNA, nucleotides and enzymes) or can be further derivatized, as shown in Fig. 6. The methods of preparing these derivatives will be published elsewhere (M.Wilchek and T.Miron). Some of these derivatives after coupling of ligands were used for the purification of enzymes and antibodies.

To summarize: stable, high capacity and charge-free agarose derivatives are available and can be used for affinity chromatography without complications that stem from ion exchange properties (N-substitutes isoureas) or hydrophobic interactions (hydrocarbon spacer) or both of them.

THE ISOLATION OF SPECIFIC PEPTIDES FOR PROTEIN

The chemical modification of proteins is an important tool in studying their structure-function relationship. The precise localization of the modified residue in the amino acid sequence is a prerequisite for the assessment of this relationship. The major difficulty in the localization of the modified amino acid is the isolation of the peptide which contains the modified residue. This is due mainly to the following: a) the modifying reagent often reacts with several residues in various yields and the digest of the protein will therefore contain several modified peptides, each present in less than molar amounts; b) the conventional methods for isolation of peptides require laborious and time-consuming steps, and each may cause a reduction in the final yield of the peptide.

Therefore, it is desirable to have a method of one-step isolation of the modified peptide from the peptide mixture derived from a modified protein. Such a method must rely on an agent which possesses affinity only for the group introduced during the modification procedures. For the purpose of this discussion we divide the modification of proteins into three categories: a) site-directed modification or affinity labelling (21) of residues at the active site of the protein; b) selective modification of one or a few residues, due to their hyperreactivity or localization at the surface of the protein (22,23,24); c) general modification of all side chains of a certain amino acid by a group specific reagent, for sequencing.

Purification of Affinity Labelled Peptides

Site-specific reagents are used extensively for affinity labelling of the active site of proteins. In these studies analogues of the substrate (in the case of enzymes) or of the haptens (in the case of antibodies) that possess a chemically reactive group are reacted with the protein. The specific and reversible binding of the analogue at the combining site of the protein is followed by the formation of a covalent bond, whereby a residue at or near that site is labelled.

We have developed a general method for the isolation of such peptides by affinity chromatography where the labelled peptide could be isolated in essentially one step (25) (Fig. 8). The method is based on the fact that the native protein (enzyme or antibody) can be used to bind, specifically, the labelled peptide by virtue of the ligand which labels it. When the native protein is covalently attached to Sepharose the ligand-coupled peptide is the only one in the digest which is bound to the protein-Sepharose column. After washing the column, the labelled peptide may be eluted under conditions

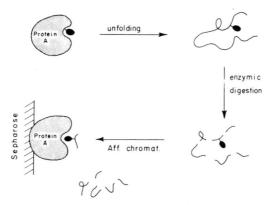

Fig. 8. A general scheme for the isolation of labelled peptides from affinity labelled protein.

that dissociate the protein ligand bond. This method was used for
the isolation of labelled peptides from affinity labelled staphylococcal nuclease or nuclease-Sepharose column (26). Analysis of
the isolated peptides indicated that Lys 48 and 49 were labelled
with a bromoacetyl analogue of thymidine 3'5'-diphosphate, while
Tyr 115 was labelled with a diazonium salt analogue.

Similarly peptides from affinity labelled anti-dinitrophenyl
(DNP) antibodies and from DNP binding mouse myeloma protein MOPC
315 labelled with N α-bromoacetyl-ε-DNP-lysine were isolated on an
anti-DNP-Sepharose column (25,27). An illustration of this approach
is described by the isolation of the labelled peptide from bovine
ribonuclease which was modified with the diazonium salt of 5'-(4-
aminophenylphosphoryl)-uridine-2'(3')-phosphate (25,28). The modified
protein was reduced and carboxymethylated. The alkylated protein
was digested with trypsin. The digest was brought to pH 5.0 (conditions under which the binding of the inhibitor to RNase is maximum),
and was applied to an RNase-Sepharose column. Fig. 9 demonstrates
that no peptide labelled with the diazo affinity reagent (absorption
at 342 nm) emerged from the column, whereas most of the digest passed
through the column. The yellow labelled peptide which was bound to
the column was eluted with 0.8 M NH_4OH as a sharp yellow band. The
yield of the isolated peptide thus obtained was more than 60% from
the amount applied to the column as judged from the absorbance at
342 nm. After one paper electrophoresis the labelled peptide was
freed from minor impurities. Amino acid analysis of this peptide
showed that it is identical to the same peptide previously isolated
by several steps of chromatography and electrophoresis. The composition of the peptide isolated was consistent with residues 67-85 in
the sequence of RNase and Tyr 73 was found to be (Fig. 10) modified.
Thus the peptide containing the modified residue was isolated
practically in one step.

Purification of Selective Modified Peptides

The method described above, can be extended to many other chemical
modifications of proteins, if specific antibodies are used as the
isolating agent. Specific antibodies with high affinity can be
elicited against almost any small molecule; they are suitable reagents for the isolation of modified peptides which have these small
molecules attached to them. In fact any chemical modification creates
a new antigenic determinant (hapten) which can raise antibodies
specific to this hapten. Peptides which contain such small molecules
as a result of chemical modification of the protein can be isolated
by such antibodies (29,30).

Classical methods for separation and isolation of peptides are
generally based upon certain physical properties (charge, size,
hydrophobicity, conformation, etc.) which vary quantitatively from
one peptide to another. In contrast, this method takes advantage

PREPARATION OF SPACER CONTAINING DERIVATIVES 25

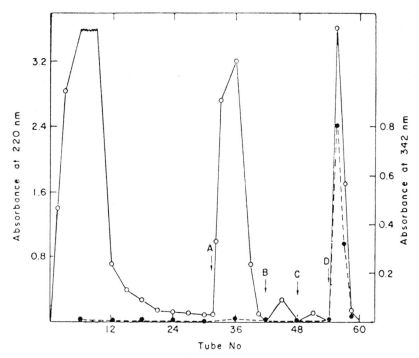

Fig. 9. Adsorption and elution of PUDP-peptide on RNase-Sepharose columns. A - Wash with 0.02 M Na acetate pH 5.0. B - Wash with 0.05 M Na acetate. C - Wash with 0.1 M Na acetate. D - Elution with 0.8 M NH_4OH.

Fig. 10. Structural formula of RNase and peptides isolated by affinity chromatography. See arrows.

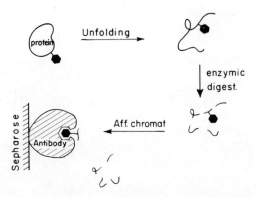

Fig. 11. A general scheme for the isolation of peptides from modified proteins on antibody-Sepharose column.

of the biological specificity of antibodies. It is based on the affinity of the antibody to a certain ligand attached to the peptide and is independent of the physical properties of the peptide itself. Only peptides which possess such a ligand will be specifically sequestered by the antibodies and thereby removed from the entire peptide mixture (Fig. 11).

The following examples will illustrate the use of this approach (29). Mono-DNP-ribonuclease was prepared according to Murdock et al. (30). The preparation contained 0.9 dinitrophenyl group per mole of RNase. Performic acid oxidized DNP-ribonuclease was digested with trypsin and chymotrypsin. The yellow digest was applied to an anti-DNP-Sepharose column. The elution pattern of this column is given in Fig. 12. All the yellow material was adsorbed by the column and 85% of the original absorbance at 360 nm could be eluted by 6 M guanidine HCl. After this treatment the antibody-Sepharose conjugate lost about 20% of its potency. Amino acid analysis of this peptide corresponds to residues 40-46 in ribonuclease (Fig.10).

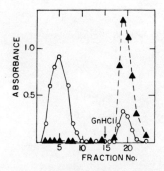

Fig. 12. Adsorption and elution of DNP-lysine peptide from RNase on anti-DNP antibody-Sepharose column.

Application of Affinity Chromatography to Protein Sequence Studies

The determination of the amino acid sequence of a given protein provides vital information in correlating protein structure with function. The classical method of peptide sequence determination is very complicated. The advent of the protein sequenator helped to a great extent in alleviating part of the problems involved, in that the process of Edman degradation can now be automated. This improvement in amino acid and sequence analysis has made peptide purification the rate-limiting step in the sequencing of proteins. The existing methods of separation and identification of peptides resulting from partial hydrolysis are tedious. The combination of electrophoresis and chromatography is not directly selective, since the basis of separation is generally peptide charge, mobility and size. A complex chromatographic pattern usually emerges - especially in the case of large molecular weight proteins - and subsequent isolation often involves successive steps which tend to affect the purity and yield.

A better method of peptide separation would result if one would recognize specific peptides biologically according to their amino acid content. This can be done by utilizing chemical modification in conjunction with affinity chromatography. If a specific modification of all the residues of the same kind (e.g. tyrosine, histidine or tryptophan etc.) can be performed, one can foresee the use of insoluble antibodies for the isolation of all peptides containing these residues.

Affinity chromatography enables the isolation of minute amounts of proteins such as receptors, transport proteins, etc. Methods for isolating peptides from digests of these proteins for sequencing must therefore be more sensitive than those afforded by today's techniques. The specific modification of the different residues with fluorescence or radioactive haptenic groups will enable the isolation of very small amounts of peptides on antibody columns for sequencing.

Anti-DNP antibodies have been used for the isolation of peptides containing residues to which a DNP group has been covalently linked by various means. After partial hydrolysis, peptides containing the DNP modified amino acid residues can thus be detained on an anti-DNP antibody column. The advantages of using DNP groups are: a) it has a high molar extinction coefficient and therefore is monitored easily; b) anti-DNP antibodies can easily be prepared; c) radioactive reagents containing DNP groups can be easily prepared.

Table II illustrates some of the reactions currently under study for the modification of different residues with the introduction of a DNP group. Tryptophan residues in proteins can be modified with the highly specific reagent 2,4-dinitrophenyl-sulphenyl chloride

TABLE II

THE USE OF ANTI-DNP ANTIBODIES FOR THE ISOLATION OF PEPTIDES

Amino acid	Reagent RX	Condition	Products
Cysteine	NO_2-C$_6$H$_3$(NO_2)-NH(CH$_2$)$_4$CH(NHCOCH$_2$Br)-COOH NO_2-C$_6$H$_3$(NO_2)-F	pH 7.5, 3 hr, 25°C pH 5, 1 hr, 25°C	-NH-CH-CO-NH- | CH$_2$ | S | R
Methionine	NO_2-C$_6$H$_3$(NO_2)-NH(CH$_2$)$_4$CH(NHCOCH$_2$Br)-COOH	24 hrs, 25 8 M urea pH 3.5	NH-CH-CO-NH | CH$_2$ | CH$_2$ | $\oplus$S - R | CH$_3$
Tryptophan	NO_2-C$_6$H$_3$(NO_2)-SCl	50% acetic acid 1 hr, 25°C	-NH-CH-CO-NH | CH$_2$ | (indole)-R
Histidine	NO_2-C$_6$H$_3$(NO_2)-NH(CH$_2$)$_4$CH(NHCOCH$_2$I)-COOH	pH 5, 24 hrs 25°C	NH-CH-CO-NH- | CH$_2$ | (imidazole: RN, NH)
Lysine	NO_2-C$_6$H$_3$(NO_2)-SO$_3$Na	pH 5, 24 hrs 25°C	NH-CH-CO-NH | (CH$_2$)$_4$ | NH | R

(31,32). The tryptophan residues are converted by this reaction into a DNP-thioether function in the two positions of the indole nucleus, and can be selectively adsorbed and purified on an anti-DNP antibody column. The method was used for the isolation of the tryptophan containing peptides from human serum albumin and from cytochrome C (32, 33).

Cysteine containing peptides can be isolated on anti-DNP columns after modification with N^{α}-bromoacetyl-N^{ϵ}-DNP-Lysine or with fluorodinitrobenzene. This approach was used for the isolation of Cys 25 containing peptides from papain (T.Miron and M.Wilchek).

Of particular interest are modifications of methionine residues for the purpose of obtaining overlapping peptides to establish positions of CNBr cleavage. The CNBr cleavage will give all the peptides containing C-terminal methionine while alkylation with N -bromoacetyl-N -DNP-lysine (BADL) will give peptides with methionine in the middle of the peptide chain. The two methionine residues in lysozyme were alkylated with BADL. The labelled peptides are now being isolated (M.Wilchek and T.Miron).

CONCLUDING REMARKS

The introduction of Sepharose as a convenient support for insolubilized antibodies and the improvement of various techniques for affinity chromatography facilitate rapid and easy procedures for the isolation of modified peptides. The high capacity of CNBr-activated Sepharose and the use of purified antibodies allows the employment of very small columns for the isolation of micromole amounts of the modified peptide. The specific isolation of modified peptides by adsorption to and elution from an antibody column can be rapidly accomplished in 1-2 hrs.

The adsorption of most (about 95%) of the modified peptides from the peptide mixture can be accomplished under mild conditions (neutral pH and various ionic strengths). The elution of the modified peptides from the antibody-Sepharose column, however, requires different conditions depending on the ligand as well as on the antibodies. It is found that 1 M NH_4OH, if it dissociates the antibody hapten bonds, is the best solvent for elution of adsorbed peptides. After this treatment the column can be regenerated to almost full capacity and can be re-used many times. In addition, almost no antibodies are released from the Sepharose (less than 0.1% of the bound antibodies) and the eluted peptides are obtained pure.

On the other hand, the use of 6 M guanidine HCl required for elution of DNP-peptides causes substantial inactivation of the antibodies and releases some 2-3% of the antibody which is bound covalently to the Sepharose. Attempts should be made to prepare antibodies to coloured charged molecules which can be eluted under mild conditions.

Obviously, if the eluate contains several modified peptides, conventional methods should be used to further separate and purify them. Extension of this concept to other haptens would result in a powerful tool for peptide separation prior to sequence determination.

REFERENCES

1. CUATRECASAS,P., WILCHEK, M. & ANFINSEN, C.B. *Proc. Nat. Acad. Sci.U.S.A. 61*:636 (1968).
2. WILCHEK, M. *Biochem.J. 127*; 7p (1972).
3. WILCHEK, M. & GIVOL,D. "Peptides 1971", North-Holland Publishing Co., 203 (1973).
4. CUATRECASAS, P. ¢ WILCHEK, M. *Biochem. Biophys. Res. Commun. 33*; 225 (1968).
5. CUATRECASAS, P. *J. Biol. Chem. 245*:3059 (1970).
6. WILCHEK, M.,& ROTMAN, M. *Israel J. Chem. 8*:172p (1970).
7. WALSH, D.A., PERKINS, J.P. & KREBS, E.G. *J.Biol.Chem. 243*:3763 (1968).
8. CORBIN, J.D., REIMANN, E.M., WALSH, D.A. & KREBS, E.G. *J. Biol. Chem. 245*:4849 (1970).
9. WILCHEK, M., SALOMON, Y., LOWE, M. & SELINGER, Z. *Biochem.Biophys. Res. Commun. 45*:1177 (1971).
10. TESSER, G.I., FISCH, H.U. & SCHWYZER, R. *FEBS Letters 23*:56 (1972).
11. SVENSSON, B. *FEBS Letters 29*:167 (1973).
12. AXEN, R. & ERNBACK, S. *Eur. J. Biochem. 18*:351 (1971).
13. WILCHEK, M. *FEBS Letters 33*:70 (1973).
14. WILCHEK, M. Proceedings of the Ninth International Congress of Biochemistry. Stockholm, p.9 (1973).
15. ER-EL, Z., ZAIDENZAIG, Y. & SHALTIEL, S. *Biochem. Biophys. Res. Commun. 49*:383 (1972).
16. SHALTIEL, S. & ER-EL, Z. *Proc. Nat. Acad. Sci. U.S.A. 70*:778 (1973).
17. HOFSTEE, B.H.J. *Biochem. Biophys. Res. Commun. 50*:751 (1973).
18. LAMED, R., LEVIN, Y. & WILCHEK, M. *Biochim. Biophys. Acta 304*: 231 (1973).
19. WILCHEK, M. & LAMED, R. Methods in Enzymology, 1974. Volume edited by W.Jakoby and M. Wilchek (in press).
20. LAMED, R., LEVIN, Y. & OPLATKA, A. *Biochim. Biophys. Acta 303*: 163 (1973).
21. SINGER, S.J. *Adv. Prot. Chem. 22*:1 (1967).
22. VALLEE, B.L. & RIORDAN, J.F. *Ann. Rev. Biochem. 38*:733 (1969).
23. SHAW, E. *Physiol. Rev. 50*:244 (1970).
24. COHEN, L. *Ann. Rev. Biochem. 37* (1968).
25. GIVOL, D., WEINSTEIN, Y., GORECKI, M. & WILCHEK, M. *Biochem. Biophys. Res. Commun. 38*:827 (1970).
26. WILCHEK, M. *FEBS Letters, 7*:161 (1970).

27. GIVOL, D., STRAUSBAUCH, P.H., HURWITZ, E., WILCHEK, M., HAIMOVICH, J. & EISEN, M.N. *Biochemistry 10*:3461 (1971).
28. GORECKI, M., WILCHEK, M. & PATCHORNIK, A. *Israel J.Chem. 8*: 173p (1970).
29. WILCHEK, M., BOCCHINI, V., BECKER, M. & GIVOL, D. *Biochemistry 10*:2828 (1971).
30. MURDOCK, A.L., GRIST, K.L. & HIRS, C.H.W. *Arch.Biochem. Biophys. 114*:375 (1966).
31. SCOFFONE, E., FONTANA, A. & ROCCHI, R. *Biochemistry 7*:971 (1968).
32. WILCHEK, M. & MIRON, T. *Biochim.Biophys. Acta 278*:1 (1972).
33. WILCHEK, M. & MIRON, T. *Biochem. Biophys. Res. Commun. 47*:1015 (1972).

QUANTITATIVE PARAMETERS IN AFFINITY CHROMATOGRAPHY

A. H. Nishikawa, P. Bailon, & A. H. Ramel

Chemical Research Division
Hoffmann-La Roche Inc.
Nutley, N. J., 07110

While the notion of purifying biopolymers in general (and proteins in particular) by selectively adsorbing them onto a specific carrier was reported some time ago (Lerman, 1953), a great surge in the application of this technique has recently followed the introduction of the term "affinity chromatography" (Cuatrecasas, et al. 1968). While mnemonically useful, this term does not adequately describe many of the processes reported in papers on affinity chromatography. Most papers typically describe a solid-phase adsorption process using a specifically designed carrier material. Rarely is any description made of relative mobilities, of partition coefficients, or of separation factors. Indeed a column is not even required as demonstrated by Porath (1972) and others.

We have begun a study of specific binding and avidity (affinity), issues which are common to all of the reported works on the affinity adsorption process. Recent papers have pointed to the importance of proper ligand concentration within an affinity sorbent to effect successful binding and separation from contaminants (Hixson & Nishikawa, 1973; Robert-Gero & Waller, 1972; Rosenberry et al, 1972). To assess the effect of ligand concentration in a gel, it is useful to consider the adsorption isotherm as a possible model.

Adsorption theory.

A kinetic approach to derive the binding equation is shown in Figure 1. The soluble enzyme [E] reacts with the insoluble ligand [L] to form the insoluble complex [EL]. At equilibrium, the forward reaction (with the rate constant, k_1) will of course be equal to the reverse reaction (with the rate constant, k_2). The

ADSORPTION ISOTHERM MODEL - DERIVATION

$$E + L \underset{k_2}{\overset{k_1}{\rightleftharpoons}} EL \qquad K_b = \frac{k_1}{k_2} = \frac{[EL]}{[E][L]}$$

at eqbrm. $k_1 [E][L] = k_2 [EL]$

total possible $[L_o] = \underset{\text{free}}{[L]} + \underset{\text{bound}}{[EL]}$

substitute for $[L]$, $k_1 [E][L_o] - k_1 [E][EL] = k_2 [EL]$

regroup, $[EL] k_2 + [EL] k_1 [E] = k_1 [E][L_o]$

and $[EL] = \dfrac{k_1 [E][L_o]}{k_2 + k_1 [E]} = \dfrac{K_b [E][L_o]}{1 + K_b [E]}$

Fig. 1

total possible binding sites, $[L_o]$ is the sum of the unbound ligands $[L]$ and those occupied by enzyme $[EL]$. After substituting the expression for $[L]$, regrouping of terms, and dividing through by the quantity: $k_2 + k_1 [E]$, we obtain the final expression for $[EL]$. By replacement of the rate constants with the equilibrium binding constant K_b, we obtain an expression whose form is exactly that for the Langmuir adsorption isotherm: $x/M = aC/1 + bC$. The complex $[EL]$ is in units of moles of enzyme per unit volume of gel, and is clearly limited by the binding site concentration $[L_o]$ designed into the gel.

Assuming a modest binding constant of 1000 for a hypothetical system, we have plotted $[EL]$ as a function of $[E]$ for three fixed values of $[L_o]$ as seen in Figure 2. By plotting $[E]$ on a logarithmic scale, we observed the expected sigmoidal pattern of binding. As expected too, the saturation limits approach that of the $[L_o]$ for each of the examples. We see also that the 50% saturation point in each case would be at about 1 mM enzyme.

It is important to consider all of the assumptions made with this simple model as listed in Figure 3.

1. <u>Ligand in gel behaves like freely soluble molecule</u>. In reality at least one degree of freedom in translational entropy is lost.

2. <u>Ligand concentration in gel, [L], is closely approximated by its measured concentration per unit volume of settled gel</u>. For the time being this is the only practical measure.

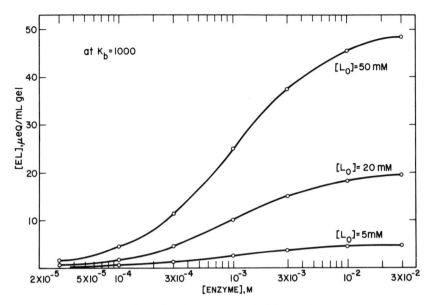

Fig. 2 Ideal Enzyme Binding Plot

ASSUMPTIONS OF ADSORPTION MODEL

1) LIGAND IN GEL BEHAVES LIKE FREELY SOLUBLE MOLECULE.
2) LIGAND CONCENTRATION IN GEL, $[L_0]$, IS CLOSELY APPROXIMATED BY ITS MEASURED CONCENTRATION PER UNIT VOLUME OF SETTLED GEL.
3) ENZYME FREELY INTERACTS WITH ALL ACCESSIBLE LIGANDS. INACCESSIBLE LIGANDS, HAVE NO EFFECT ON BINDING POTENTIAL OF ENZYME.
4) CARRIER PRESENTS NO SPECIAL EFFECTS ON ENZYME-LIGAND BINDING, OTHER THAN STERIC OCCLUSION OF SOME LIGANDS.

Fig. 3

3. <u>Enzyme freely interacts with all accessible ligands. Inaccessible ligands have no effect on binding potential of enzyme.</u> Obviously here we ignore steric limitations of tunnels, pockets, etc., on enzyme access.

4. <u>Carrier presents no special effects on enzyme-ligand binding other than steric occlusion of some ligands.</u> The truly inert carrier does not yet exist, ionic problems are frequently manifest.

Such an idealized model may be considered the simplest of all possible worlds and would only serve as a starting point in the detailed understanding of affinity binding.

Experimental.

Affinity system preparation. Because of prior experience as well as the large body of available data, the affinity purification system for trypsin was used in our model studies. The details for preparation of the specific affinity gel are shown in Figure 4.

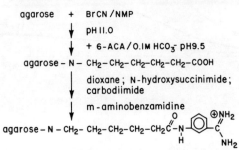

Fig. 4

Beads of agarose were treated with a solution of cyanogen bromide dissolved in N-methylpyrrolidone (NMP). While the gel suspension was stirred and adjusted to pH 11 in a jacketed vessel at 4 - 10°C, the BrCN solution was added slowly so that the in situ temperature was held within these limits. After collecting and washing with 0.1 M bicarbonate buffer at pH 9.5, the beads were then treated with 6-aminocaproic (6-ACA) acid in the same buffer.

The 6-ACA-agarose was collected washed with water then with dioxane. This was then treated with N-hydroxysuccinimide (NHS) and 1-ethyl-3-(3'-dimethylamino-propyl-) carbodiimide hydrochloride. The active ester 6-ACA-gel thus obtained in dioxane was treated with an equal volume of m-aminobenzamidine in bicarbonate buffer at pH 8.3 for 4 hrs. at 4°C.

Equilibrium binding studies were carried out as diagrammed in Figure 5. Two ml of the affinity gel was mixed with 2 ml of enzyme

QUANTITATIVE PARAMETERS IN AFFINITY CHROMATOGRAPHY

Equilibrium binding experiments

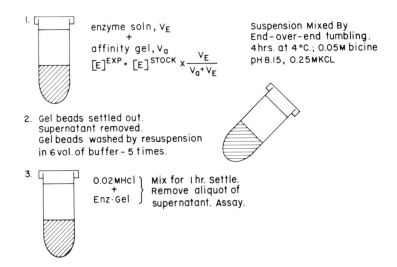

Fig. 5

solution and the suspension agitated by end-over-end tumbling in a capped tube. The mixing was done at least 4 hrs at 4°C in 0.05 M bicine buffer at pH 8.15 with 0.25 M KCl. After the binding period the gel beads were settled out and the unbound enzyme in the supernatant was drawn off. The gel was washed by resuspension in 6 volumes of the same buffer and settling out. The process was repeated four more times. Finally the EL complex was disrupted by treatment with 0.01 M HCl for 1 hr. After settling the gel beads, an aliquot of the supernatant liquid was assayed for activity.

Results and discussion.

The results of some equilibrium binding studies are shown in Figure 6. The affinity gels were all made with 6% agarose (Sepharose 6B). The 'ideal' binding curve was not seen. With the gel bearing high ligand concentration (48.9 μeq/ml gel), technical problems, for the moment precluded determination of $[E]$ at lower $[E_T]$ values. In gels with lower $[L_o]$ the apparent saturation limits only roughly correlate with the ligand concentration. More striking is the asymmetry of the sigmoidal curves. The $[EL]$ characteristically increases slowly up to the 50% mark then rises abruptly to saturation. We calculate that in the gel with highest ligand concentration, only 0.8% of the L-sites are occupied at saturation; with the low ligand gel, only 1.2% are occupied. Hence, in all of the gels

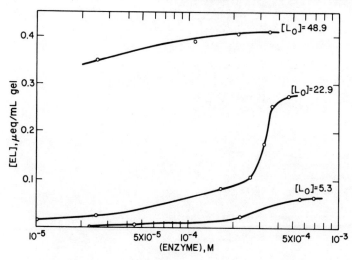

Fig. 6 Trypsin Binding to 6% Gels

only a fraction of the ligands appear to be bound by enzyme under saturation concentrations. Thus surface over-crowding by enzyme does not appear to explain the curve asymmetries. While not yet conclusive, the pattern of binding curves seems to be one where the 50% saturation point shifts upward in enzyme concentration, [E], as the concentration of L_o decreases. The significance of this trend is that the [E] corresponding to 50% saturation may be a measure of binding avidity (affinity).

Clearly the adsorption model as proposed is only an approximation. Figure 7 more graphically illustrates one of our great simplifications, namely the assumption that $[L_o]$ the ligand concentration in gel bead is the same as that measured in the settled volume beads. Because the agarose polymer chains are somewhat restrictive on enzyme movement, the thermodynamic activity coefficient for the ligand towards the enzyme active-site cannot be expected to be the same as that for the freely soluble compound. Thus an accurate measure of enzyme affinity to a ligand awaits the determination of the intra-gel ligand concentration.

A further consideration of this point is explored in Figure 8. The high ligand gel ($[L_o]$=48.9) was diluted with an equal volume of plain 6% agarose. The result was a gel with [L] = 24 µeq/ml gel bed. The binding curve reveals a saturation capacity level less than half of the undiluted gel-this may due to electrostatic effects from carrier. But the 50% binding point is still at a lower [E] than for

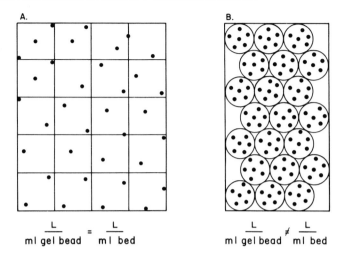

$$\frac{L}{ml\ gel\ bead} = \frac{L}{ml\ bed} \qquad \frac{L}{ml\ gel\ bead} \neq \frac{L}{ml\ bed}$$

Fig. 7

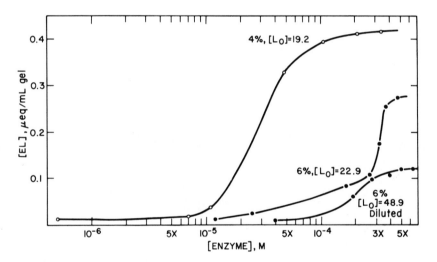

Fig. 8 Comparison of Gels in Trypsin Binding

the gel with $[L_o] = 22.9$ μeq/ml as determined by chemical treatment only. The 4% agarose gel containing L=19.2 μeq/ml was most interesting in its resemblance to ideal behavior. Here even with only a moderate ligand concentration the saturation capacity is like that of the high ligand 6% gel. The increased gel porosity no doubt accounts for the higher capacity but the affinity to the enzyme is apparently greater than the 6% gel with L=22.9 in view of the lower

[E] corresponding to 50% saturation.

In our discussion we have now alluded several times to a distinction between enzyme capacity per unit volume of gel and the avidity by which the enzyme binds. Ligand concentration within the gel appears to affect both properties but not in identical fashion. The diagrammatic representations in Figure 9 may be useful in distinguishing capacity and affinity. The picture on the left represents gel beads of high affinity and its corresponding bed of high

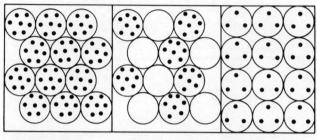

High Bed capac.   Low Bed capac.   Low Bed capac.
High [L],(affinity) High [L],(affinity) Low [L],(affinity)

Fig. 9 Enzyme Capacity vs. Affinity

capacity. If we dilute this gel with plain agarose we get the situation in the middle picture. Here the high affinity persists since the intra-gel ligand concentration is still high in the derivatized beads. But the bed capacity is lower since not all of the beads are capable of enzyme binding. Finally, the picture on the right represents a situation where we have uniformly derivatized the gel beads with low concentration of ligand. We obtain a gel with low affinity and low bed capacity.

Most of the foregoing considerations of course lead to questions on column operations with affinity gels. On Figure 10 we see the results of two column experiments. In comparing the EL complex to the ES complex in Michaelis-Menten kinetics, it seemed that the gel capacity under equilibrium saturation conditions ($[EL]^{max}$) would be higher than that attainable under column modes of operation. But we didn't know by how much. In the upper diagram of Figure 10 is the elution pattern of 151 mg trypsin dissolved in 4 ml buffer and applied to a column bed of 7 ml high affinity gel (6%, L=48.9). The amount of enzyme applied corresponded to the gel capacity under saturation conditions. We observed 75% of the enzyme to pass directly out of the gel indicating an overload. About 25% of the activity was affinity bound.

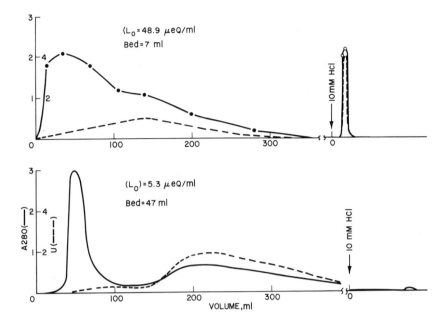

Fig. 10. Affinity Chromatography of Trypsin.

In the lower diagram the same total amount of trypsin was applied to 47 ml of a low affinity/capacity gel. The bed volume was determined by the capacity of this gel under saturation conditions -- the calculated capacity of this column bed was the same as that of the first. The elution pattern indicates that the enzyme is not bound at all. Less than 0.1% of the input was seen upon acid elution of gel. Hence, we have ligand recognition but no strong binding. But despite this some purification was seen since the bulk of the active protein had a slower mobility than the inactive protein. Here was an example of affinity chromatography in the sense of liquid chromatographers, i.e. a separation based on differential mobility.

References.

Cuatrecasas, P., Wilchek, M., and Anfinsen, C.B.(1968) Proc. Nat. Acad. Sci., USA 61, 636.
Hixson, H.F., and Nishikawa, A. H. (1973) Arch. Biochem. Biophys. 154, 501.
Lerman, L. (1953) Proc. Nat. Acad. Sci., USA 39, 233.

Porath, J., and Sundberg, L. (1972) in "The Chemistry of Biosurfaces", Edit. by M.L. Hair, Marcel Dekker, N.Y.C., Chap. 14.
Robert-Gero, M., and Waller, J. P. (1972) Eur. J. Biochem. 31, 315.
Rosenberry, T. L., Chang, H. W., and Chen, Y. T. (1972) J. Biol. Chem. 247, 1555.

NON-SPECIFIC BINDING OF PROTEINS BY SUBSTITUTED AGAROSES

B. H. J. Hofstee
With the technical assistance of N. Frank Otillio

Biochemistry Division
Palo Alto Medical Research Foundation
Palo Alto, California 94301

Previous results (1-3) showed that at slightly alkaline pH and at relatively low ionic strength (e.g., 0.01-0.05 M Tris-HCl) a large number of negatively charged but unrelated proteins invariably were strongly bound by agarose (Sepharose 4B) substituted with an n-alkylamine or with 4-phenyl-n-butylamine (PBA).[1] Unsubstituted agarose or agarose treated with CNBr, but without subsequent addition of amine, did not bind these proteins. Positively charged protein species generally showed little affinity for the substituted agaroses. Of the proteins tested, the only exception was α-chymotrypsin which, despite its positive charge, showed strong affinity for the PBA-substituted material. By contrast, strong binding did not occur when the ligand was an n-alkylamine.

The binding of chymotrypsin by agarose-bound PBA probably is a specific event such as that between enzymes and their substrates or substrate analogues. Specific affinity, whether or not involving an "active" site, presumably depends to a large extent on complementarity of the contours (i.e., fit) of the interacting molecules. In any event, in the absence of a specific effect, binding to the amine-substituted agaroses is greatly affected by

---

[1] The abbreviations used are: PBA, 4-phenyl-n-butylamine; BSA, bovine serum albumin; β-LG, β-lactoglobulin; OV, ovalbumin; γ-G, 7S γ-globulin; EG, ethylene glycol; DMF, dimethylformamide; C_n, alkylamine with n straight-chain C-atoms; A-C_n, agarose substituted with C_n.

the overall charge of the protein. The observed preference for binding of negatively charged protein species implicates the positive charge on the agarose-bound amino group (see ref. 4) as playing a predominant role in the binding. However, there is evidence (1,2) that the binding also may be enhanced by the hydrophobic moiety of the ligand, possibly due to a mutually reinforcing effect of the (long range) electrostatic and (short range) hydrophobic forces. In fact, the binding is often virtually "irreversible," i.e., without increasing the ionic strength and/or decreasing the polarity of the eluant, no detectable amount of protein is released from its complex with the adsorbent. These observations are further substantiated by the present results.

On the one hand, nonspecific protein binding interferes with affinity chromatography of the type whereby the ligand is designed to selectively bind a particular protein in a mixture. On the other hand, this type of binding presents many opportunities for the separation of proteins by a more general type of affinity chromatography. Chromatography with the present adsorbents, carrying hydrophobic groups in addition to charges, may have advantages over procedures based on charge effects alone (see ref. 2). Therefore, the need for further study of the nonspecific type of binding is two-fold. First, such studies may reveal how to prevent this type of binding in order to improve chromatographic procedures based on specific affinity and second, they may show how to improve its application to a general "hydrophobic" chromatographic procedure as an adjunct to other nonspecific procedures, i.e., those based on molecular size or on electric charge alone.

EXPERIMENTAL

Materials

Bovine serum albumin (BSA), β-lactoglobulin (β-LG), ovalbumin (OV) and 7S γ-globulin (γ-G) were the same highly purified commercial preparations used previously (1,2). Sepharose 4B (lot 5979) was obtained from Pharmacia and CNBr as well as 4-phenyl-n-butylamine (PBA) from Aldrich. The n-alkylamines were purchased from Eastman-Kodak. The amines were recrystallized as the hydrochlorides from ethanol through the addition of ether at $\simeq -15°$. Ponceau S was from Harleco and ethylene glycol (EG) was 'Baker analyzed.'

The Sepharose derivatives were prepared by the general procedure described by Cuatrecasas (5). Coupling of the amines with the CNBr-activated material was carried out at pH 10.2 in the presence of 50 per cent dimethylformamide (DMF) in water. Since the amines buffer in the applied pH region, no further buffer was added. The final product was washed exhaustively with 50 per cent (v/v) DMF and/or 50 per cent EG (v/v) in the presence of M NaCl.

Methods

The degrees of substitution of the adsorbents were determined from the capacities for "irreversible" binding of ovalbumin or of Ponceau S, as described under RESULTS. A titrimetric method for this purpose also is described in that section.

The relative degree of the polar (e.g., electrostatic) or apolar (hydrophobic) aspects of the binding of a protein by an adsorbent was determined by the effects of the addition to the eluant of salt or of ethylene glycol, respectively (1-3). For several experiments, particularly those in which elution was carried out by means of a salt gradient, an automated procedure was used whereby the eluant was pumped into the column with a peristaltic pump and the protein content of the eluate was monitored continuously through measurement of the U.V. light absorbance at either 280 or 225 nm. However, stepwise elution with manual measurement of absorbance of the eluate has the advantage over automated procedures in that a large number of columns can be operated simultaneously. Details of the procedures are described in the legends to the Figures and the Table.

RESULTS

Apparent Inhomogeneity and "Irreversibility" of the Binding of Proteins by the Adsorbents

Earlier observations (1-3) indicated that the binding of negatively charged proteins by the present cationic adsorbents often is "irreversible." At least part of the protein remains bound after exhaustive washing with the applied buffer solution (0.01-0.05 M Tris-HCl). Additional studies of this phenomenon are shown in Fig. 1. For these studies an 0.1 per cent solution of ovalbumin in 0.001 M Tris-HCl buffer, pH 8, was applied to a 2 ml cooled ($\simeq$ 5°) column of the n-butylamine-substituted adsorbent (equilibrated with the buffer) until the column was saturated with protein. The loaded column contained about 30 mg of protein of which roughly one third could be washed off with the buffer. However, almost 20 mg remained bound after exhaustive washing, unless the ionic strength of the eluant was raised. This observation suggests inhomogeneity with respect to binding of the protein to the adsorbent. Inhomogeneity also is indicated when the bound protein is eluted by a salt gradient (Fig. 1A) and is even further emphasized when the elution is carried out by increasing the ionic strength stepwise (Fig. 1B). The data indicate that at each level of the ionic strength a fraction of the protein is released while the remainder is "irreversibly" bound under the ambient conditions, i.e., it cannot be eluted without changing the composition of the eluant.

Although even highly purified proteins often appear inhomo-

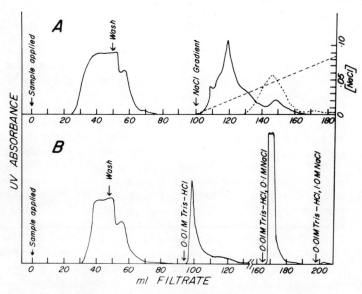

Figure 1

Apparent inhomogeneity and "irreversibility" of the binding of ovalbumin by n-butylamine-substituted Sepharose 4B. The protein, in a concentration of 1 mg/ml in 0.001 M Tris-HCl, pH 8, was applied to a 2 ml cooled ($\simeq 5°$) column of the adsorbent until the latter was saturated, i.e., until the light absorbance (A_{280}, solid curves) of the filtrate reached a constant value. The column then was washed with the buffer until no further protein was released. The loading, washing and subsequent elution was done continuously by means of a peristaltic pump and the adsorbance of the filtrate was continuously monitored. Part A, the loaded and washed column was eluted with the aid of an NaCl-gradient in the buffer. The dotted curve was obtained when only 2 mg of the protein (instead of a saturating amount) was applied to the column and the eluate was monitored at 225 instead of 280 nm (see legend to Fig. 4). Part B, the elution was carried out by a stepwise increase of the ionic strength.

geneous by isoelectric focusing (see ref. 6), the gross inhomogeneities observed here more likely are caused by inhomogeneity of the binding sites of the adsorbent than by that of the protein. This is further confirmed by results obtained under the same conditions but with 2 mg (instead of $\simeq$ 20 mg) of protein bound to the column (see dotted curve of Fig. 1A). It can be seen, that the elution of the smaller amount of protein (monitored at 225 instead of 280 nm) occurs at salt concentrations at which the last part of the larger amount is eluted. The results suggest that certain "strong" binding sites are occupied first by the protein and that several other sites of decreasing affinity become occupied when more protein is applied to the column. This interpretation could also apply to previous observations (2) showing that in some cases (e.g., serum albumin bound by an adsorbent with a strongly hydrophobic group, see also below), only part of the protein can be eluted by salt (NaCl) in concentrations as high as 1 M, whereas the remainder is dislodged by the addition to the eluant of a polarity reducing agent such as ethylene glycol (see DISCUSSION).

Irreversible Binding of Dye (Ponceau S). Determination of the Degree of Substitution of the Adsorbents

Ponceau S, which carries hydrophobic groups in conjunction with an overall negative charge, is bound by the amino-agaroses in an "irreversible" fashion, similar to the case of many proteins. Under a given set of conditions and after application of a saturating amount of the dye, a certain amount of the dye remains bound after washing with the solvent even after the filtrate becomes colorless. However, removal of the "excess" (i.e., relatively weakly bound) dye proceeds much more slowly in 0.01 M Tris-HCl buffer, pH 8, than when the column is equilibrated, loaded and washed in the presence of water alone. The amount of dye "irreversibly" bound under the ambient conditions, but eluted with 1 M n-octylamine-HCl in 50 per cent ethylene glycol (see also ref. 3), was about the same in both cases. Therefore, the Ponceau-binding capacity of the adsorbents usually was carried out in water without buffer.

As before (3), the ratios of "irreversibly" bound ovalbumin and of Ponceau S dye were about the same for n-butylamine- and n-caprylamine-substituted agaroses. The data of Fig. 2A show that this ratio is constant for all of the adsorbents tested. The ratio of the binding capacities of the protein and the dye is about 9:1 on a weight basis. On a molar basis there is more than 5 times as much dye as ovalbumin. The constant ratio is evidence that the same groups of the adsorbent are involved in the binding of the protein and of the dye and that the extent of binding is a measure of the degree of substitution.[2] This is confirmed by the

---

2)
 Although the data of Fig. 2 might suggest an increase in the degree of substitution with decreasing C-chain length of the ligand, such a conclusion would not be justified since the experimental conditions of the substitution process, including the extent of activation of the Sepharose, were not necessarily identical for all preparations (see also footnote to page 8).

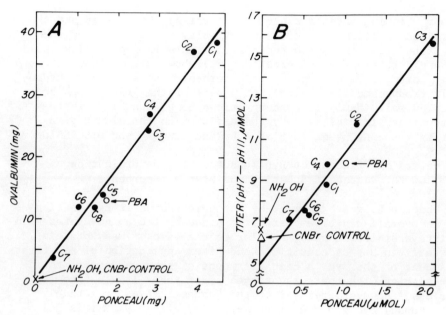

Figure 2

Part A. Relationship between the binding capacities of substituted agaroses for "irreversible" binding of ovalbumin or Ponceau S. Saturating amounts of the protein or of the dye were applied to 1 ml columns of agarose substituted with n-alkyl-amines of varying C-chain lengths (C_1, C_2, etc.), or substituted with 4-phenyl-n-butylamine (PBA), or were applied to columns of hydroxylamine (NH_2OH)-treated and of untreated (but aged) CNBr-activated agarose (CNBr control). The columns then were washed with the ambient solvent (0.01 M Tris-HCl for the protein and water for the dye, see text) until the light absorbance (at 280 nm for the protein and 525 nm for the dye) reached zero. The amounts that were bound were determined from the light absorbances after elution of the protein by 1 M NaCl or of the dye by 1 M n-octylamine-HCl in 50 per cent (v/v) ethylene glycol.

Part B. Relationship between acid-base titer and Ponceau binding capacity per ml of the same adsorbents as in part A, although most often with a lower degree of substitution (see text). The adsorbents (7-15 ml of settled slurry in a total volume of 25 ml of 1.0 M KCl) were titrated from pH 7 to pH 11 with KOH. Ponceau binding and elution was carried out as for Part A.

absence of binding of either protein or dye by CNBr-activated but unsubstituted agarose (Figs. 2A and 2B).

As seen in Fig. 2B, an estimation of the relative degree of substitution also can be obtained from the acid-base titer of the adsorbent. The amino-groups of the ligands retain their basic properties after binding to agarose (4) and the substituted adsorbent thus should show an increase in acid-base titer in the alkaline region. The data of Fig. 2B indicate this and suggest that the increase in titer is largely superimposed on the already considerable titer of the CNBr-control. On the other hand, little or no additional titer is produced when the freshly CNBr-activated Sepharose is treated with hydroxylamine suggesting that little or no reaction occurs or that the product is unstable (see below).

Rate of decrease in the degree of substitution of the adsorbents upon storage. The degree of substitution of the adsorbents often decreased gradually upon storage, even when refrigerated (see also ref. 7). For instance, in one case a decrease of as much as 85 per cent in the Ponceau binding capacity had occurred during a period of almost 5 months. In another case the decrease was about 40 per cent in 40 days. However, the treatment of the adsorbents was not always the same. Data on the stability of a number of adsorbents under more specified conditions are presented in Table I. These data suggest that the adsorbents with the highest degree of substitution are the least stable. On the other hand, a decrease in stability with decreasing C-chain length also could be involved.

TABLE I

Effect of storage on the Ponceau S binding capacities of Sepharose 4B substituted with n-alkylamines of varying C-chain lengths (C_2, C_4, ...) or with 4-phenyl-n-butylamine (PBA). The dye-binding capacities (mg dye per ml slurry) were measured (see text) before and after storage of the adsorbents for 17 days at $\simeq$ 4° in water containing 1.0 M KCl.

	C_2	C_4	C_6	C_7	C_8	PBA
Initial	.875	.61	.41	.27	.19	.78
After 17 days	.47	.47	.39	.26	.20	.40
Decrease	.405	.14	.02	.01	0	.38
Per cent	46	23	$\simeq$ 5	$\simeq$ 0	$\simeq$ 0	49

Effect of Relatively Small Hydrophobic Groups on the Electrostatic Binding of Proteins by the Adsorbents

The predominant role of electrostatic forces in the binding of protein by n-alkylamino-agaroses is confirmed by the observation (Fig. 3) that the binding, at least to the adsorbents with the smaller hydrophobic groups (C_1-C_6), is readily reversed by salt. The C-chain length has little effect on the salt concentration at which the proteins are released in a salt gradient (Fig. 4). The smaller hydrophobic groups do not form hydrophobic "bonds" as is indicated by the similar behavior of proteins with greatly different hydrophobic properties, e.g., ovalbumin and serum albumin (8,1,2), on the C_1 and the C_4-adsorbents (Fig. 4). One might conclude, therefore, that these adsorbents act merely as ion exchangers. However, the data of Fig. 2 show that CNBr-treated but unsubstituted agarose, which seems to carry positive charges and at extremely low ionic strength binds negatively charged proteins (1), does not bind these proteins in 0.01 M Tris-HCl buffer. Treatment of freshly activated Sepharose with hydroxylamine does not improve its protein binding capacity, but treatment of the activated agarose with an amine carrying even the smallest hydrophobic group (methylamine) results in an effective adsorbent (Figs. 2 and 3).[3] Presumably, the relatively smaller hydrophobic groups of the ligands alone are ineffective in protein binding, but may shield the positive charges on the adsorbent-bound amino groups from the quenching effect of water (see DISCUSSION).

Binding of Proteins by Immobilized Ligands with C-chain Lengths > C_6

It is apparent from Fig. 3 and Fig. 5 that a different and more direct hydrophobic effect is involved in protein binding by adsorbents carrying ligands with C-chain lengths > C_6. This is indicated by the increasing stability of the binding in 1 M NaCl with increasing C-chain length of the ligand and by a similar increase in binding with increasing hydrophobicity of the protein. Fig. 5 shows the effect of the C-chain length of a homologous series of n-alkylamino-agaroses on the extent of binding of various proteins in the presence of 1 M NaCl. Strong binding at this high ionic strength increases sharply when the C-chain length is

---

[3) This observation seemingly contradicts previous results (1) indicating the absence of strong binding of these proteins by methylamine-substituted agarose. These results were obtained, however, with adsorbents prepared from an earlier lot (nr. 2700) of Sepharose 4B, the degrees of substitution of which were generally much lower than those of the present lot (nr. 5979), in particular in the case of methylamine.

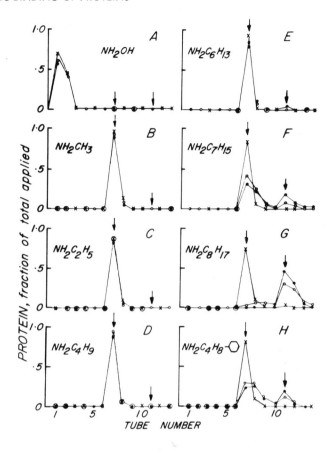

Figure 3

Effects of salt and of ethylene glycol on the fractional binding of bovine serum albumin (solid circles), β-lactoglobulin (open circles) and of ovalbumin (crosses) by substituted agaroses. The adsorbents were prepared by treatment of CNBr-activated agarose (Sepharose 4B) with an excess of hydroxylamine (part A), n-alkylamines of increasing C-chain lengths (parts B-G) or of 4-phenyl-n-butylamine (part H). One ml, 8 mm wide columns of these materials were equilibrated at 0-5° with 0.01 M Tris-HCl buffer, pH 8, charged with 1-2 mg of protein in buffer and washed with five 2 ml portions of buffer. Subsequently, four 2 ml portions of 1.0 M NaCl in buffer were applied (starting at the first arrow) followed by the application (starting at the second arrow) of several 2 ml portions of buffer containing 1.0 M NaCl as well as 50 per cent (v/v) ethylene glycol. The amounts of protein in the tubes were determined from the light absorbance at 280 nm. The results with n-propylamine- or n-valerylamine-substituted agaroses (not shown) were similar to those of parts B, C and D.

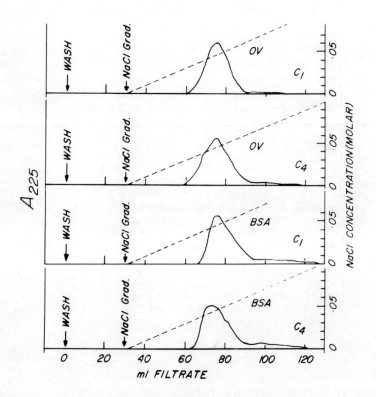

Figure 4

Effect of salt (NaCl) concentration on the binding of ovalbumin (OV) and of bovine serum albumin (BSA) by agarose substituted with methyl(C_1)- or n-butyl(C_4)-amine. Two mg of a protein was applied to a 2 ml cooled ($\simeq 5°$) column of an adsorbent under otherwise the same conditions as for Fig. 1A. Elution was carried out with a linear salt (NaCl) gradient and the eluate was continuously monitored by the light absorbance at 225 nm.

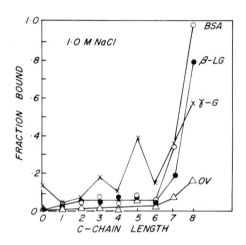

Figure 5

Effect of the ligand C-chain length of n-alkylamino-agaroses on their binding of serum albumin (BSA), β-lactoglobulin (β-LG), ovalbumin (OV) or 7S γ-globulin (γ-G) in 0.01 M Tris-HCl, pH 8, containing 1.0 M NaCl. Two mg of a protein was applied to a 1 ml column of an adsorbent equilibrated with the buffer alone. The fraction bound refers to the protein not eluted by six 2 ml portions of buffer followed by four 2 ml portions of the NaCl-buffer mixture (see Fig. 3).

increased to > C_6. With the C_8-adsorbent the extent of binding relates to the protein as BSA > β-LG > γ-G > OV, which is similar to the relative degrees of hydrophobicity of these proteins as determined by binding of a naphthalene derivative (8).

As compared to the C_2-, C_4- and C_6-adsorbents, γ-globulin shows a relatively high affinity for A-C_3 and A-C_5. It would appear that a specific effect involving oscillation between odd and even numbered C-chains is operative here, similar to the effect of the substrate C-chain length on the activity of certain enzymes (9).

Hydrophobic aspects of the apparent inhomogeneity with respect to protein binding by the adsorbents. Apparent inhomogeneity of certain highly purified proteins (e.g., BSA and β-LG) with respect to chromatography on adsorbents of the present type was first observed with PBA-agarose (2). Only part of BSA adsorbed on a column of this material could be eluted by salt (NaCl) at concentrations higher than 1 M, whereas elution of the remainder required the addition of a polarity reducing agent such as ethylene glycol. It appears from the present observations and previous data (1,2) that this type of inhomogeneity depends on the hydrophobicity of the protein and of the adsorbent. The binding is largely or completely reversible by 1 M NaCl if either the adsorbent or the protein or both are of low hydrophobicity. This applies to all three proteins of Fig. 3 chromatographed on adsorbents with C-chain lengths < C_6 and for ovalbumin even when the C-chain length of the adsorbent is > C_6. On the other hand, when both reactants are strongly hydrophobic, e.g., BSA and the C_8-adsorbent (Fig. 3G), all of the protein remains bound in the presence of M NaCl. It is only in intermediate cases that this type of inhomogeneity is observed, e.g., when BSA is chromatographed on A-C_7 or A-PBA[4] (Fig. 3F and 3H, respectively). It would seem that in such intermediate cases the two types of binding, i.e., the predominantly electrostatic and the more hydrophobic types, occur simultaneously (see DISCUSSION).

The fraction of BSA that remains bound to A-PBA in 1 M NaCl is increased by decreasing the load of the column (Fig. 6A). This is consistent with the above assumption that the apparent inhomogeneity is not necessarily caused by inhomogeneity of the protein but rather by the binding sites of the adsorbent not being identical. When a larger amount of protein is applied to the column, the fraction of protein molecules for which strong binding sites would be available decreases.

---

4)
It has been shown previously (see ref. 17) that the hydrophobicity of a benzene ring corresponds to 3-4 straight chain C-atoms. Thus the -PBA ligand would be less hydrophobic than -C_8 (see also ref. 18).

Fig. 6B indicates that the chromatographic behavior of the fraction released by M NaCl and subsequently dialyzed is similar to that of the original protein. However, the fraction that remained bound in M NaCl showed a greater degree of hydrophobicity when it was rechromatographed after elution by 1 M NaCl containing 50 per cent ethylene glycol, followed by dialysis overnight against the buffer. Like the original protein it showed two fractions,

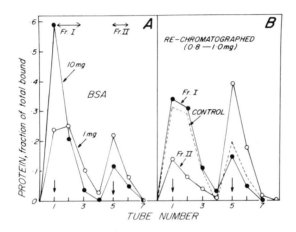

Figure 6

Fractionation of bovine serum albumin through chromatography on 4-phenylbutylamine-substituted agarose under the same conditions as for Fig. 3, i.e., after washing with the buffer alone 1 M NaCl was added starting at the first arrow and starting at the second arrow 50 per cent ethylene glycol was also present in the eluant.
Part A. Effect of the protein load (10 mg and 1 mg on a 1 ml column of the adsorbent) on the relative amounts eluted in the presence of 1 M NaCl (Fraction I) and in the added presence of 50 per cent ethylene glycol (Fraction II).
Part B. Rechromatography of 0.8-1.0 mg of the dialyzed fractions I and II of Part A, as compared to 1 mg of the unfractionated but dialyzed control (dashed curve).

but the relative amount of the second, more hydrophobic, fraction was greatly increased. Contrary to expectations, this was not the case with BSA after binding to $A-C_8$ (unpublished). Although chromatography of BSA on $A-C_8$ (Fig. 3G) suggests an even stronger hydrophobic interaction than with A-PBA, after elution by the NaCl-ethylene glycol mixture and dialysis, the chromatographic behavior on A-PBA was about the same as that of the original protein treated identically but without prior binding to $A-C_8$ (see DISCUSSION).

DISCUSSION

Protein binding in the presence of 1 M NaCl (e.g., with BSA bound to $A-C_8$, Fig. 3G) not necessarily indicates that electrostatic forces are not involved. As shown previously (2,3), in the absence of specific interaction, binding generally requires that the protein and the adsorbent are oppositely charged, even with $A-C_8$. Also, the dissociation of the complex of BSA and $A-C_8$ required an increase in the ionic strength as well as the presence of a polarity reducing agent (unpublished). In any event, it would appear that polar (electrostatic) as well as apolar (hydrophobic) forces are operative in protein binding by these adsorbents, those with the larger as well as those with the smaller hydrophobic groups. However, the latter may merely function as a shield against quenching of the electrostatic forces by water without actually forming hydrophobic "bonds" (see refs. 10,11).[5]

For the larger hydrophobic groups an additional and stronger interaction with the formation of hydrophobic bonds may be superimposed on the shielding effect. The stronger interaction does not necessarily involve "detergent" action with accompanying irreversible denaturation (see ref. 12). This is confirmed by the observation that the chromatographic behavior of BSA on A-PBA, after prior adsorption on $A-C_8$ is similar to that of the original protein. On the other hand, a less easily reversible change is observed with "Fraction II" eluted from A-PBA (see Fig. 6), suggesting a certain degree of specificity of the interaction, i.e., the change in the protein induced by the phenyl-butyl-group would be longer lasting than that by the straight-chain n-octyl group. A similar change in the apparent homogeneity of α-lactalbumin with respect to chromatography on DEAE-Sephadex or DEAE-cellulose and depending on the previous treatment of the protein, was observed by Hopper (13), who interpreted the phenomenon on the basis of

---

5) Recent results (19) suggest that this may also apply to the binding of negatively charged proteins by diethylaminoethyl (DEAE)-agarose which, similar to the present adsorbents, carries positive charges in conjunction with small hydrophobic groups.

electrostatic effects alone. Further investigation of these matters would seem of great interest.

Interaction of a protein with the substituting ligands of the adsorbent could be affected by the location of the ligands and by their concentration on the adsorbent as follows. If the concentration is such that the average distance between two carrier-bound ligand molecules is smaller than the diameter of the protein molecule, the possibility of multiple point attachment obtains and could be one of the causes of the extremely strong binding that was observed. Such strong binding is most likely to occur in a cavity on the adsorbent surface, particularly when the protein molecule would fit into the cavity. On the other hand, at points on the matrix where the bound ligands are distributed over a protruding area, binding would be less strong. Since the surface of the adsorbent may be assumed to be irregular, many different situations in addition to these two hypothetical cases, would obtain. In view of the limited number of the most favorable binding sites, the binding would become stronger as the protein:adsorbent ratio is decreased (see Fig. 1A).

When a column of the adsorbent is initially saturated with protein (Fig. 1) the results can readily be interpreted on the basis of binding sites with a range of affinities from extremely weak to extremely strong. Release of the protein from the weak binding sites can be achieved merely by washing with a low ionic strength buffer. With continued washing the rate of release would gradually decrease and eventually approach zero, since further release would become more and more difficult without changing the composition of the eluant. When actual hydrophobic binding also is involved, release of the most strongly held fraction of the protein may require a decrease in the polarity of the eluant as well as an increase in the ionic strength.

Small amounts of protein bind more homogeneously on a particular column than a large amount (Fig. 1A). This could mean that reducing the amount of applied protein causes one type of site on the adsorbent to predominate in the binding. The virtual immobilization of several enzymes on this type of adsorbent (3) probably is related to the relatively small amounts that were applied, i.e., all of the enzyme was bound to strong binding sites.

If multiple point attachment were one of the reasons for strong nonspecific binding, one way to diminish it would be to lower the degree of substitution to the point where the distance between the substituting groups is larger than the diameter of the protein molecule (14). This would not affect the specific "one-to-one" interaction such as that between an enzyme active site and an immobilized substrate analogue. The present and previous results (1,2) indicate that in order to further diminish non-specific interaction, it would be expedient to attach the ligand to an uncharged "arm" that offers no opportunity for strong hydrophobic binding. A charge effect alone, even in conjunction with weak hydrophobic interaction, can be counteracted at high ionic strength (see Fig. 3). On the other hand, for the present type of adsorbent the positive charge of the substituting amine can be turned into an asset. All of the negatively charged protein species in a mixture may first be removed by a positively charged adsorbent that has no specific affinity for the protein of interest. The remaining positively charged species would generally be electrostatically repelled by a positively charged adsorbent. However, with a properly designed ligand, a particular positively charged protein may be specifically bound despite a positive charge of the adsorbent, e.g., the binding of (positively charged) α-chymotrypsin by agarose substituted with ε-amino-n-caproyl-D-tryptophan methylester (15), or by 4-phenyl-n-butylamine (16).

SUMMARY

The present results with agaroses substituted with n-alkylamines of varying C-chain length ($C_1 - C_8$), further substantiate the virtual "irreversibility" of nonspecific binding of many proteins by adsorbents carrying hydrophobic groups in conjunction with charges. For all of the present adsorbents the apparent irreversibility of the binding of negatively charged proteins seems to depend on the positive charge of the agarose-bound amino group as well as on the hydrophobic moiety. With bovine serum albumin an actual hydrophobic "bond," stable at high salt concentrations (1-4 M NaCl) appears to be formed only when the C-chain length of the substituting ligand is $> C_6$. With a less hydrophobic protein, such as ovalbumin, even the binding to the C_8-ligand is readily reversed by 1 M NaCl. However, elution of ovalbumin from the C_4-adsorbent by a stepwise increase in the ionic strength (I) indicates that at a particular level of I only part of the protein can be eluted, the rest remaining "irreversibly" bound, until the I of the eluant is raised. Elution by means of a salt gradient also shows inhomogeneity. The evidence suggests that this inhomogeneity is related to inhomogeneity of the binding sites of the adsorbent rather than to that of the protein.

ACKNOWLEDGMENTS

This work has been supported by U.S. Public Health Service Grant No. FR-05513, the Harvey Bassett Clarke Foundation and by the Santa Clara County United Fund.

REFERENCES

1. Hofstee, B.H.J., Bioch. Biophys. Res. Comm. 50, 751 (1973).
2. Hofstee, B.H.J., Anal. Bioch. 52, 430 (1973).
3. Hofstee, B.H.J., Bioch. Biophys. Res. Comm. 53, 1137 (1973).
4. Cuatrecasas, P., and Anfinsen, C. B., in "Methods in Enzymology," Vol. XXII (W. J. Jacoby, Ed.) p. 351, Academic Press, New York (1971).
5. Cuatrecasas, P., J. Biol. Chem. 245, 3059 (1970).
6. Bobb, D., Ann. N.Y. Ac. Sci. 209, 225 (1973).
7. Cuatrecasas, P. and Parikh, I., Biochemistry 11, 2291 (1972).
8. McClure, W. O. and Edelman, G. M., Biochemistry 5, 1908 (1966).
9. Hofstee, B.H.J., J. Biol. Chem. 199, 365 (1952).
10. Jencks, W. P., "Catalysis in Chemistry and Enzymology," McGraw Hill, 1969.
11. Epstein, H. F., J. Theor. Biol. 31, 69 (1971).
12. Reynolds, J. A., Herbert, S., Polet, H. and Steinhardt, J., Biochemistry 6, 937 (1967).
13. Hopper, K. E., Bioch. Biophys. Acta 293, 364 (1973).
14. Schmidt, J. and Raftery, M. A., Biochemistry 12, 852 (1973).
15. Cuatrecasas, P., Wilcheck, M. and Anfinsen, C. B., Proc. Nat. Ac. Sci. USA 61, 636 (1968).
16. Stevenson, K. J. and Landman, A., Can. J. Biochem. 49, 119 (1971).
17. Hofstee, B.H.J., Arch. Bioch. Biophys. 78, 188 (1958).
18. Tanford, C., J. Am. Chem. Soc. 84, 4240 (1962).
19. Hofstee, B.H.J., Polymer Preprints, in press.

A SOLID PHASE RADIOIMMUNE ASSAY FOR ORNITHINE TRANSCARBAMYLASE

Donald L. Eshenbaugh, Donald Sens and Eric James

Department of Chemistry

University of South Carolina, Columbia, South Carolina
29208

Radioimmune assays utilizing solid supports were introduced by Catt et al (1966) and Wide and Porath (1966). The classical radioimmune assay described by Yalow and Berson (1960) is based on the ability of unlabeled antigen to compete with antigen labeled with a radioactive isotope bound to antibody. After permitting the system to reach equilibrium antibody bound radioactivity was separated from unbound radioactivity and one or the other determined in order to ascertain the quantity of unlabeled antigen in the unknown sample.

Such a competitive radioimmune assay is not applicable to the study of regulation of ornithine transcarbamylase biosynthesis, performed in vitro, due to the complexity of the arginine biosynthetic regulon. The eight enzymes comprising this pathway are coded by nine genes (Gorini, Gundersen and Burger, 1961) scattered around the E. coli chromosome, as shown in fig. I. There are two genes, argI and argF which code for ornithine transcarbamylase resulting in the formation of four multimolecular forms of the enzyme (Legrain, Halleux, Stalon and Glansdorff, 1972). Strains of E. coli K12 are available which have argF deleted; however, there is no known deletion covering argI and available point mutations are leaky. The enzymes of the arginine biosynthetic regulon are controlled in a coordinate but nonparallel manner (Gorini et al 1961, Glansdorff, Sand and Verhoef, 1967, Cunin, Elseviers, Sand, Freundlich and Glansdorff, 1969) and the repressed, derepressed ratio of argF is higher than that of any other enzyme of the pathway rendering this the enzyme of choice for developing studies of regulation, performed in vitro.

We describe a solid phase radioimmune assay for ornithine

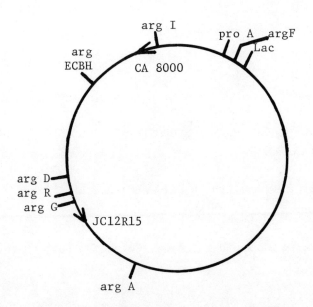

Figure 1. Simplified genetic map of E. coli

Table I

Bacterial Strains

Strain	Genotype	Origin
Hfr strains		
CA8000	Hfr H thi	Beckwith
JC12R15	thi met purC argR15 spc^r lac xyl mtl	Jacoby
F$^-$ strains		
GL5	thi(lac pro argF)$^\nabla$ argI^{am}	Glansdorff
GL2	thi leu pro argF purA argI	Glansdorff
EJ106	thi(lac pro argF)$^\nabla$ argR15 spc^r argI^{am}	GL5 X JC12R15
EJ107	thi argR15 spc^r argI^{am}	EJ106 X CA8000

transcarbamylase which may be used for the rapid and specific determination of ornithine transcarbamylase synthesized in a coupled transcription-translation system, directed by the argF gene carried on the specialized transducing bacteriophage $\lambda h80C_I 857$ dargF isolated by James and Gorini, (in preparation).

MATERIALS AND METHODS

A. Materials

Sepharose 4B, DEAE Sephadex A-50 and Sephadex G-200 were obtained from Pharmacia. Freund's adjuvant was purchased from GIBCO. Carbamyl phosphate, lysozyme, and glass beads were purchased from Sigma Chemical Co. Deoxyribonuclease and L-ornithine were obtained from General Biochemicals. Tryptone and yeast extract were purchased from Difco Laboratories. Polyethylene glycol 6000 was purchased from J. T. Baker. Chloramine T was obtained from Aldrich Chemical Co. Ammonium sulfate was of special enzyme grade obtained from Schwarz Mann. Sodium iodide-125 was purchased from New England Nuclear. Immuno-diffusion plates were purchased from Hyland Division Travenol Laboratories. Materials for polyacrylamide disc gel electrophoresis and 2,3-butanedione monoxime were obtained from Eastman Organic Chemicals. All other chemicals were of reagent grade. Supplies for liquid scintillation counting were purchased from Beckman Instruments.

B. Strains

The genotype and origin of all strains used in this work are listed in Table 1 and the locations of the relevant markers on the genetic map of E. coli K12 are shown in Figure 1.

C. Media and Genetic Procedures

Selection plates contained medium A (Davis and Mingioli, 1950), supplemental growth factors as required, 2% agar and 0.5% glucose as carbon source. Supplements were used at the following concentrations (in µg/ml): L-arginine, 100; L-citrulline, 100; L-leucine, 80; L-ornithine, 100; L-proline, 50; thiamine, 10; L-threonine, 50.

For selection against streptomycin sensitive strains, streptomycin sulfate (500 µg/ml) was added to minimal or rich medium. L medium (Lennox, 1955) was used for growth of bacteria. Conjugation experiments were performed as described by James and

Gorini (in preparation) and transductions with P1 were performed using the procedure of Lennox (1955).

D. Enzyme Assays

Ornithine transcarbamylase activity was determined as described by Gorini (1958) using 0.5 ml of diluted enzyme solution pre-warmed to 37°C followed by the addition of 0.5 ml of assay mixture (0.1 ml of 1M tris-HCl, pH 7.8, 0.025 ml of 0.2 M $MgCl_2$, 0.375 ml of 1% L-ornithine and 3.52 mg of carbamyl phosphate). The mixture was incubated at 37°C for ten minutes and the reaction terminated by the addition of 5 ml of 0.05 M HCl followed by boiling for ten minutes. The solution was cooled and 2 ml of acid 6 (0.5 ml of concentrated sulfuric acid and 1.5 ml of 85% phosphoric acid) and 0.2 ml of aqueous 2,3-butanedione monoxime (375 mg per ml) were added. The solution was mixed thoroughly and transferred to a boiling water bath for thirty minutes. After cooling to room temperature the absorbance of the samples was determined at 470 nm and correlated with a standard curve for citrulline concentration. One unit of OTCase converts one micromole of ornithine to citrulline per hour. Arginosuccinase activity was determined as described by Jacoby and Gorini (1967).

E. Growth of Cells

One hundred liters of EJ107 were grown at 37°C with aeration in L medium with glucose. Cells were harvested in late log phase by sedimentation for 16 hours at 4°C in the presence of polyethylene glycol and 0.5 M NaCl (Eshenbaugh, Sens and James, 1974) yielding approximately 300 grams of cells.

F. Preparation of Ornithine Transcarbamylase

Three hundred grams of cells were suspended in 300 ml of 0.05 M phosphate buffer, pH 7.5, warmed to 37°C and three hundred mg of lysozyme was added. The suspension was stirred slowly at 37°C for 15 minutes; two mg of deoxyribonuclease was added and the suspension was stirred for an additional 5 minutes at 37°C. This mixture was then transferred to a Waring blender, cooled to 4°C, and nine hundred grams of finely divided glass beads were added followed by blending for 20 minutes at medium speed. The suspension was allowed to settle, the supernatant decanted and the residue of glass beads was blended with an additional 300 ml of 0.050 M phosphate buffer for five minutes. The supernatants from these extractions were combined and centrifuged for 15 minutes at

6000 rpm in a Sorvall GSA rotor, followed by centrifugation at 30,000 x g for 2 hours in a Beckman L5-75 ultracentrifuge to remove remaining cell debris.

Heat denaturation in the presence of substrate: The 30,000 x g supernatant was adjusted to 10 mM L-ornithine and heated for 5 minutes in 75 ml aliquots in a 2 l baffle bottom flask in a New Brunswick gyrorotary shaker bath maintained at 65°C. The mixture was rapidly cooled in an ice water bath and centrifuged for 5 minutes at 6000 rpm in a Sorvall GSA rotor.

Ammonium sulfate fractionation: The heat denatured fraction was adjusted to 50% ammonium sulfate saturation by slowly adding 30 grams of solid ammonium sulfate per 100 ml, allowed to equilibrate for 30 minutes at 4°C followed by centrifugation for 30 minutes at 12,000 rpm. The 50% supernatant was then adjusted to 80% ammonium sulfate saturation by the slow addition of 20 grams of ammonium sulfate per 100 ml and allowed to equilibrate for 30 minutes at 4°C followed by centrifugation as described. The 80% saturated ammonium sulfate precipitate was resuspended in a minimum amount of 0.050 M phosphate buffer, pH 7.5, and dialyzed overnight against 0.050 M phosphate buffer.

DEAE-Sephadex chromatography: The dialyzed 80% precipitate was loaded on a DEAE-Sephadex A-50 column (100 x 5 cm), equilibrated with 0.050 M phosphate buffer, pH 7.5, and washed (flow rate = 30 ml/hr) with the same buffer until no further protein was eluted. OTCase was eluted with a KCl gradient from 0 to 0.3 M KCl in 50 mM phosphate, pH 7.5. OTCase elutes at approximately 0.1 M KCl in 50 mM phosphate pH 7.5. Fractions containing activity were amalgamated.

Gel filtration on Sephadex G-200: The enzyme solution was dialyzed against 50 mM phosphate buffer, pH 7.5, and concentrated by ultrafiltration to about 10 ml and applied to a Sephadex G-200 column (2.5 x 85 cm) equilibrated with 50 mM phosphate, pH 7.5. Elution was performed by washing with the same buffer (flow rate 6 ml/hr).

Disc gel electrophoresis: Disc electrophoresis in polyacrylamide gel was performed as described by Ornstein (1964) and Davis (1964) using a refrigerated Biorad apparatus with a current of 2 ma/gel at 4-8°C.

G. Preparation and Purification of Specific Antibodies

Three rabbits were injected subcutaneously with 300 micrograms of homogeneous OTCase in Freund's complete adjuvant followed every 2 weeks by booster injections with 300 micrograms of OTCase in Freund's incomplete adjuvant. Two weeks after the second booster injection 50 ml of blood were obtained from each rabbit. Serum from each rabbit had high titres of antibody to OTCase as determined by immuno-diffusion tests using the Ouchterlony technique (Ouchterlony, 1949). Rabbits were bled every ten days, 25 ml of serum was obtained per rabbit at each bleeding and booster injections were administered every two months thereafter.

Covalent attachment of OTCase to Sepharose: Fifty ml of packed Sepharose 4B was washed with distilled water and resuspended in 50 ml of distilled water. Five grams of CNBr were dissolved in 100 ml of water. The CNBr solution was added to the Sepharose slurry with stirring; the pH was maintained at between 10.9 and 11.1 by the dropwise addition of 4N NaOH, and the temperature held at about 5°C by constant cooling in an icewater bath. After about 10 minutes activation was complete and the activated Sepharose was rapidly washed on an extra coarse glass funnel with one liter of ice cold 0.2 M phosphate buffer, pH 7.2. The activated Sepharose was then mixed with 10 mg of homogeneous OTCase and mixed slowly on a New Brunswick rotor cell unit at a speed of 1 rpm for 24 hours at 4°C. The Sepharose linked OTCase was washed twice with 100 ml 0.5 M NaCl and with two 100 ml aliquots of 0.050 M phosphate, 0.15 M NaCl, pH 7.2.

Purification of antibody by affinity chromatography: A column of Sepharose linked OTCase was poured (2 x 7 cm) and equilibrated at 4°C with 50 mM phosphate, 0.15 M NaCl, pH 7.2. Ten ml of serum were applied to the column and washed (20 ml/hr) with the same buffer. The column effluent was monitored at 280 nm using a Chromatronix dual wavelength absorbance monitor and after the absorbance returned to base line, the column was washed with 3.0 M NaSCN until the absorbance again returned to base line.

Covalent attachment of purified antibody to Sepharose: Five ml of packed Sepharose 4B was activated as described and washed with 100 ml of 0.1 M $NaHCO_3$ buffer, pH 9.0. Activated Sepharose was mixed with 7 mg of antibody dissolved in 0.1 M $NaHCO_3$ buffer, mixed for 24 hours at 4°C and washed as described above.

H. Iodination of OTCase

Iodination was performed in 50 mM phosphate buffer, pH 7.0, as described by Campbell, Garvey, Cremer and Sussdorf (1970). One mg of OTCase was added to 0.5 mCi NaI^{125} followed by 0.2 mg of freshly dissolved chloramine T in a total volume of 0.9 ml. The solution was stirred for 5 minutes and 0.2 mg of sodium metabisulfite in a volume of 0.2 ml was added to stop the oxidation by chloramine T. Unreacted iodine was removed by exhaustive dialysis against 50 mM phosphate buffer, pH 7.0.

I. Radioimmune Assay

Sepharose linked antibody (0.1 ml) was added to the solution to be assayed in a total volume of 0.5 ml, adjusted to 0.050 M phosphate, 0.15 M NaCl, pH 7.2. The assay mixture was mixed with continuous gentle rotation on a New Brunswick rotor drum at 5 rpm in a 33° incubator for 60 minutes and then diluted with 5 ml of assay buffer, centrifuged at 500 x g for 3 minutes and the supernatant discarded. The Sepharose was transferred to a Whatman GF/C 25 mm filter and washed with 50 ml of assay buffer. Radioactivity was determined using Beckman Ready-Solv VI Scintillation fluid in a Beckman LS 230 liquid scintillation counter.

RESULTS AND DISCUSSION

In order to facilitate the preparation of argF ornithine transcarbamylase, a strain was constructed which carried mutations rendering the argR and argI gene products inactive. This was achieved as outlined in Table I.

The strain GL5 produces no argF OTCase and only traces of argI OTCase due to leakiness of the amber mutation; other enzymes in the arginine biosynthetic pathway are produced at repressed levels when the strain is grown in the presence of arginine. The strain was made constitutive for enzymes in the arginine pathway by crossing in the argR15 allele with JC12R15 and crossing out the lac pro argF deletion by selection of pro^+ recombinants in an EJ106xCA8000 conjugation experiment. All pro^+ recombinants were also $argF^-$ and lac^-. The strain produced, EJ107, was shown to still carry $argI^{am}$ by transduction of GL2 utilizing P1 propagated on EJ107, and demonstrating that more than 25% of arg^+ recombinants were pro^+ and none were ade^+. Constitutive levels of arginosuccinase and OTCase are made by the strain when grown in the presence of arginine.

The strain EJ107 was grown overnight at 37°C in 2 l of L medium with 0.05% glucose and used as inoculum for 100 l of the same medium. The use of glucose starvation in the inoculum culture reduced the lag phase when the 100 l fermenter was inoculated and permitted rapid initiation of cell growth. Cells were harvested in late log phase using the procedure of rapid sedimentation in the presence of polyethylene glycol at 1 x g, (Eshenbaugh, Sens and James, 1974) thus obviating the necessity of centrifuging large volumes of cells in a continuous flow centrifuge. Approximately 1.5% of the cellular protein is

Table II

Purification of Ornithine Transcarbamylase

Fraction	Volume (ml)	Total Enzymes Activity x 10^6	Protein (mg)	Specific Activity	Purification (fold)	Yield (%)
I Crude Extract	1330	9.9	31,000	320	1.0	100
II S30	1210	9.4	23,000	410	1.3	95
III Heat Denaturation	1250	7.5	13,000	580	1.8	77
IV Ammonium Sulfate Fraction	220	7.4	3,400	2,200	7	75
V DEAE Sephadex Eluate	55	5.5	290	19,100	60	56
VI Sephadex G200 Eluate	25	4.0	90	44,000	138	41

OTCase in a constitutive strain such as EJ107 and the subsequent purification of OTCase by procedures based on those described by Legrain et al (1972) was facile, yielding a homogeneous protein with a specific activity of more than 40,000 (Table II) which exhibits one protein band when examined by disc electrophoresis in polyacrylamide gel (Figure 2).

Figure 2. Disc electrophoresis in polyacrylamide gel of argF ornithine transcarbamylase at pH 9.3.

Homogeneous OTCase was covalently linked to cyanogen bromide activated Sepharose in 0.20 M phosphate buffer, pH 7.2. These conditions resulted in the covalent binding of 100% of applied protein and retention of between 65-75% of enzymic activity. Such high retention of enzymic activity was probably obtained because of the interaction of subunits comprising the quaternary structure of the enzyme; and furthermore, the high molecular weight of ornithine transcarbamylase may have also contributed to the excellent retention of enzymic activity.

A column of Sepharose-OTCase was utilized in order to purify antibody directed specifically against OTCase. Rabbit serum was applied directly to the Sepharose-OTCase column, previously equilibrated with buffer; most of the applied protein washed through the column (Figure 3) and subsequent rapid washing with 0.1 M HCl, 1.0 M HCl, 0.1 M NH_4OH or 1.0 M NH_4OH, directly into buffer followed by immediate neutralization resulted in the elution of a protein fraction which exhibited specific immunological reaction with OTCase (data not shown); however, yields were about 20-30% as judged by quantitative precipitin tests. The use of chaotropic ions for the disruption of antigen-antibody binding has been described by Dandliker, Alonso, de Saussure, Kierszenbaum, Levison and Schapiro (1967) and by de Saussure and Dandliker (1969). Elution with 3.0M sodium thiocyanate was found to yield two neighboring protein peaks (Figure 3). Both peaks of protein exhibited specific immunological reaction with homogeneous OTCase as would be expected from the purification technique utilized in their isolation. The resolution of two antibody fractions with slightly different binding affinity, as judged by differential elution by a chaotropic ion is not unexpected as different popu-

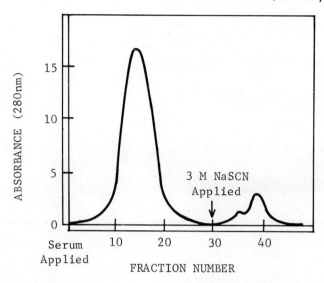

Figure 3. Affinity Chromatography of Rabbit Serum on Sepharose-OTCase.

lations of antibody may be present even in an individual animal. The fractions corresponding to these antibodies were combined and dialyzed against 0.050 M phosphate buffer, 0.15 M sodium chloride, pH 7.2. A small amount of precipitate occurred during the course of dialysis and was removed by centrifugation at 6000 x g for 20 minutes. The fraction thus obtained exhibited specific anti OTCase activity and the yield was approximately 60% of that in the applied serum. A competitive radioimmune assay as described by Yalow and Berson, (1960) is not applicable to the problem of detecting small quantities of ornithine transcarbamylase synthesized in vitro, directed by the argF gene carried on the specialized transducing bacteriophage λh80C$_I$857dargF, because large amounts of cross reacting material are present in the S30 cell extract utilized as the source of ribosomes, factors and enzymes. This is because there is no known deletion of the argI gene in E. coli K12. An amber mutation in argI was utilized, but these mutations are known

A SOLID PHASE RADIOIMMUNE ASSAY

to be leaky (Gorini, personal communication). A direct radio-immune assay is, therefore, appropriate for this problem and the use of a solid support for carrying the specific antibody was expected to be efficacious as it should obviate the necessity of operating at the equivalence point. Furthermore, it should permit rapid assays as only a primary reaction between antigen and antibody is required, whereas with precipitin reactions antigen and antibody must be incubated, at equivalence, for many hours in order to obtain a satisfactory precipitate. A solid phase antibody-antigen complex sediments rapidly at low g forces and may be recovered and counted efficiently on glass fiber filters.

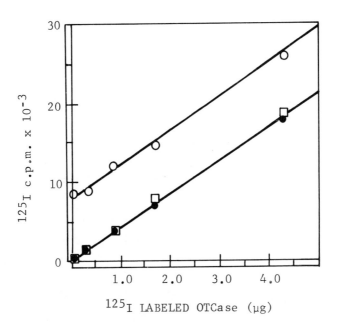

Figure 4. Relation between ^{125}I radioactivity specifically absorbed by Sepharose-AntiOTCase and quantity of antigen present in assay. Antigen along , Antigen in the presence of 650 µg unlabeled S30 , Antigen in the presence of 650 µg ^{125}I labeled S30 (specific activity 780 cpm/µg) .

Although experiments employing the assay would involve the detection of ^{3}H labeled OTCase synthesized in vitro, it is difficult to obtain ^{3}H labeled OTCase of high specific activity in vivo. To alleviate this problem, ^{125}I labeled OTCase was made to test the assay system, realizing that ^{3}H OTCase synthesized in vitro would have a high specific activity (Austin and McGeoch, 1973) permitting the detection of very small quantities of newly synthesized ^{3}H labeled OTCase. The utility of the procedure is shown in Figure 4 in which a linear relationship between ^{125}I cpm and quantity of antigen is observed both for antibody alone and in the presence of a large excess (650 µg) of unlabeled S30 protein. This clearly indicates that the amount of cross reacting and contaminating protein present in an in vitro protein synthesis system will not interfere with the assay. The data showing the relationship of ^{125}I cpm and antigen in the presence of 650 µg of ^{125}I labeled S30 indicates that approximately 10 µg of cross reacting material are present in 650 µg of S30 having a specific activity of 780 cpm/µg. This cross reacting material presumably arises from the leakiness of the argI^{am} mutation and also from the partial polypeptide chain produced from it.

The radioimmune assay described provides a rapid, sensitive and specific assay for OTCase and is being utilized in studies aimed at elucidating the control mechanism of the arginine biosynthetic pathway, performed in vitro.

ACKNOWLEDGEMENT

We thank Drs. Behnke, Gorini, Glansdorff, Jacoby, Mizejewski, and Wolf for advice and for sending us strains; Miss Diane Winburn, Mr. Philip James, and Mrs. Glenda Pickler for excellent and patient technical assistance. This work was supported by the University of South Carolina Productive Scholarship fund, the Research Corporation Brown-Hazen-BH-853 grant and American Cancer Society Grant VC-131, all provided to Eric James.

REFERENCES

1. Austin, S. and McGeoch, D., PNAS, 70, 2420 (1973).
2. Baumberg, S. and Ashcroft, E., J. Gen. Microbiol., 69, 365 (1971).
3. Campbell, D. H., Garvey, J. S., Cremer, N. E. and Sussdorf, D. H., Methods in Immunology (second edition), 149, W. A. Benjamin, Inc., New York (1970).
4. Catt, K. J., Niell, H. D. and Tregear, G. W., Biochem. J., 100, 316 (1966).
5. Cunin, R., Elseviers, D., Sand, G., Freundlich, G. and Glansdorff, N., Mol. Gen. Genetics, 106, 32 (1969).

6. Davis, B. D. and Mingioli, E. S., J. Bact., 60, 17 (1950).
7. Davis, B. J., Ann. N. Y. Acad. Sci., 121, (2), 404 (1964).
8. Dandliker, W. B., Alonso, R., de Saussure, V. A., Kierszenbaum, F., Levison, S. A. and Schapiro, H. C., Biochemistry, 6, 1460 (1967).
9. de Saussure, V. A. and Dandliker, W. B., Immunochemistry, 6, 77 (1969).
10. Eshenbaugh, D. L., Sens, D. A. and James, E., Anal. Biochem., (in press).
11. Glansdorff, N., Sand, G. and Verhoef, C., Mutation Res., 4, 743 (1967).
12. Gorini, L., Bull. Soc. Chim. Biol., 40, 1939 (1958).
13. Gorini, L., Gundersen, W. and Burger, M., Cold Spring Harbor Symposium Quant. Biol., 26, 173 (1961).
14. Jacoby, G. A. and Gorini, L., J. Mol. Biol., 24, 41 (1967).
15. James, E. and Gorini, L., (in preparation).
16. Legrain, C., Halleux, P., Stalon, V. and Glansdorff, N., Eur. J. Biochem., 27, 93 (1972).
17. Lennox, E. S., Virology, 1, 190 (1955).
18. Ornstein, L., Ann. N. Y. Acad. Sci., 121, (2), 321 (1964).
19. Ouchterlony, O., Acta Pathol. Microbiol. Scand., 26, 507 (1949).
20. Wide, L. and Porath, J., Biochem. Biophys. Acta., 130, 257 (1966).
21. Yalow, R. S. and Berson, S. A., J. Clin. Invest., 39, 1157 (1960).

PURIFICATION OF ACETYLCHOLINESTERASE BY COVALENT

AFFINITY CHROMATOGRAPHY

Houston F. Voss, Y. Ashani,* and Irwin B. Wilson

Department of Chemistry, University of Colorado
Boulder, Colorado 80302
Israel Institute for Biological Research
Ness-Ziona, Israel*

Although the concept of affinity chromatography for the purification of macromolecules goes back many years, it has been only in the last five years or so that the field has blossomed. This development has been fostered by the creation and availability of numerous types of gels for chromatography and by the efforts of a number of highly skillful and imaginative researchers.(1) The standard type of affinity chromatography involves ligands that reversibly bind to the protein with a high degree of specificity. If the protein is an enzyme these ligands are usually selected from reversible inhibitors.

In the case of some enzymes reasonably specific covalent (irreversible) inhibitors are available and when incorporated into gel could be used for trapping the enzyme. If there is also a means of releasing trapped enzyme it might be possible to develop a purification process based upon covalent affinity chromatography.

In this paper we report on the purification and some of the properties of molecular forms of acetylcholinesterase from the electric eel. We used covalent affinity chromatography based upon the "nerve gas" reaction of acetylcholinesterase with organophosphonates that contain a good leaving group.(2) The reaction catalyzed by acetylcholinesterase

$$(CH_3)_3\overset{+}{N}C_2H_4-OCCH_3 + H_2O \rightleftharpoons (CH_3)_3\overset{+}{N}C_2H_4OH + CH_3COOH$$

occurs enzymically in two steps

$$(CH_3)_3\overset{+}{N}C_2H_4OCCH_3 + E: \rightleftharpoons CH_3\overset{O}{\underset{\|}{C}}-E + (CH_3)_3\overset{+}{N}C_2H_4OH$$

$$\Updownarrow + H_2O$$

$$CH_3COH + E$$

In the first step the enzyme serves as a remarkable nucleophile and in the second step the enzyme serves as an excellent leaving group. The nucleophilic group is the hydroxyl group of a specific serine residue.

The "nerve gas" reaction illustrated with isopropyl methylphosphonofluoridate (sarin, GB) is

$$\underset{iPrO}{\overset{CH_3}{>}}\overset{O}{\underset{\|}{P}}-F + E: \rightleftharpoons \underset{iPrO}{\overset{CH_3}{>}}\overset{O}{\underset{\|}{P}}-E + F^-$$

Here too the enzyme serves as a potent nucleophile with the nucleophilic group again the hydroxyl group of the same serine residue. In this case the enzyme is only a very poor leaving group so that the inhibition is said to be irreversible.

The specific reaction on which we have based our covalent affinity column

$$\underset{CH_3CNHC_2H_4O}{\overset{O\|}{}}\underset{}{\overset{CH_3}{>}}\overset{O}{\underset{\|}{P}}-O\Phi NO_2 + E: \rightleftharpoons \underset{CH_3CNHC_2H_4O}{\overset{O\|}{}}\underset{}{\overset{CH_3}{>}}\overset{O}{\underset{\|}{P}}-E + HO\Phi NO_2$$

occurs very readily but involves a ligand, 2 aminoethyl,p-nitrophenyl methylphosphonate,

$$\underset{NH_2C_2H_4O}{\overset{CH_3}{>}}\overset{O}{\underset{\|}{P}}-O\Phi NO_2$$

which is not a potent inhibitor of the enzyme until it becomes acylated by attachment to "extended" gel (sepharose 4B)

PURIFICATION OF ACETYLCHOLINESTERASE

Ligand

$$\text{gel} \Big) \text{-arm-} \overset{O}{\underset{\|}{C}} \text{-NH}_2\text{C}_2\text{H}_4\text{O} \diagdown \underset{\diagup}{\overset{CH_3}{\overset{\diagdown}{P}}} \overset{O}{\underset{\|}{}} \diagdown \text{O}\Phi\text{NO}_2 + \text{E} \longrightarrow$$

(1)

$$\text{gel} \Big) \text{-arm-} \overset{O}{\underset{\|}{C}} \text{-NH}_2\text{-C}_2\text{H}_4\text{O} \diagdown \underset{\diagup}{\overset{CH_3}{\overset{\diagdown}{P}}} \overset{O}{\underset{\|}{}} \text{-E} + \text{NO}_2\Phi\text{OH}$$

The arm is N-succinyl 1,5 diaminopentane. Covalent affinity columns prepared in this way contained on the order of $1\text{-}5 \times 10^{-4}$ equivalents of ligand per ml of packed column as measured by the release of p-nitrophenol with alkali. These columns have an enormous capacity to trap serine esterases. About half the ligands will react with chymotrypsin. We do not know the full capacity of the column for acetylcholinesterase but it is quite high, at least 1.2 mg/ml when the bound ligand is 5×10^{-4} M. Typically we pass 100 ml or more of a centrifuged homogenate of electric organ containing 1.5 mg of enzyme in 10% ammonium sulfate, 1 M NaCl through one ml of column at the rate of 1 ml/minute. At least 85% of the enzyme is trapped. "All" the protein comes through.

Enzyme that has been inhibited by an organophosphorus anticholinesterase can usually be reactivated by nucleophiles. These nucleophiles force the enzyme to serve as a leaving group. N-methyl pyridinium-2-aldoxime (2PAM) is especially effective;

$$\text{CH}_3\overset{O}{\underset{\|}{C}}\text{-NC}_2\text{H}_4\text{-O} \diagdown \underset{\diagup}{\overset{CH_3}{\overset{\diagdown}{P}}} \overset{O}{\underset{\|}{}} \text{-E} + \underset{\underset{CH_3}{|}}{\overset{}{N^+}}\diagdown \overset{N}{\underset{H}{C}} \diagdown \text{OH} \;\rightleftharpoons$$

$$\text{CH}_3\overset{O}{\underset{\|}{C}}\text{-NC}_2\text{H}_4\text{-O} \diagdown \underset{\diagup}{\overset{CH_3}{\overset{\diagdown}{P}}} \overset{O}{\underset{\|}{}} \text{-O} \diagdown \text{N} \overset{H}{=} \text{C} \diagdown \underset{\underset{CH_3}{|}}{N^+} + \text{E}$$

We expected to release the trapped enzyme from the column with 2PAM. However this does not occur even though 2PAM does release p-nitrophenol from the column.

We are able to use this column because of another phenomenon. Bound protein, chymotrypsin or acetylcholinesterase, is slowly released in an inhibited form at a rate of 5-30% per day. This released enzyme presumably contains gel substance.(3) Released acetylcholinesterase is readily reactivated with 2PAM.

In a typical procedure enzyme from a centrifuged homogenate is trapped as above. Then the column is washed with 50-100 ml of 1 M NaCl buffer. Finally 5 ml of buffer containing 0.01 M 2PAM is passed through the column and the column is set aside for 4 days. Two ml of 2PAM solution is then added to displace the solution containing enzyme from the column. At least 20% of the enzyme is obtained. (The process can be repeated with further recovery of enzyme - with patience, almost all the enzyme can be recovered.) Enzyme is separated from 2PAM by gel permeation with G-50 or Sepharose 6B which also separates the 11S form. The specific activity of the enzyme is 5-6 mmoles per mg per minute (Table I). The trapping of enzyme apparently occurs as envisaged. Enzyme inhibited with diethyl phosphorofluoridate which phosphorylates the active site (esteratic site) is not trapped by the column (Fig. 1).

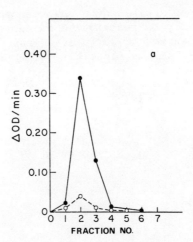

Figure 1. Elution profile of acetylchlinesterase from affinity column. ●────● Enzyme inhibited by DEFP before passage through column. ○────○ Uninhibited enzyme.

TABLE I

Sample	Volume	Activity (mmole hr^{-1} ml^{-1})	Protein (mg/ml)	S.A.	% Rec.
Homogenate	85	11.7	–	–	–
10,000 G Super.	75	11.7	6.9	0.59	–
100,000 G Super.	75	10.0	5.0	2.0	100
Passed thru column	75	2.1	5.0	0.42	20
Trapped by column	–	–	–	–	80
Released from column (60 hrs.)	2 ml	47	–	–	12.5
G-50 column	4 ml	23.0	0.065	360	12.5 (16% of trapped)

A fresh homogenate of frozen electric organ of electric eel contains 18S, 14S and a small amount of 8S enzyme forms as evaluated by sucrose gradient velocity sedimentation.(4) In our hands it contains no 11S enzyme. On standing 11S enzyme appears and finally all enzyme forms have been converted to the 11S form.(5,6) Enzyme released from the column after two days contains some 11S but after 6 days a considerable amount of 11S enzyme has appeared (Fig. 2).

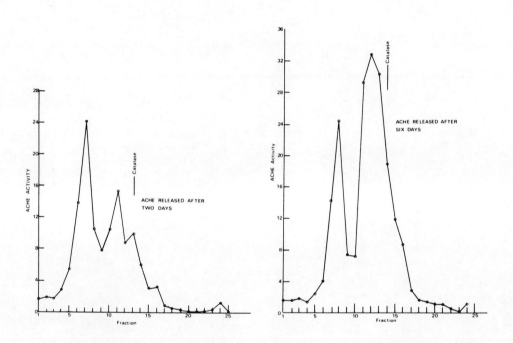

Figure 2. Sucrose gradient profiles of AChE released from affinity column at indicated times. Catalase is 11.4S.

A purified enzyme preparation was subjected to sucrose gradient centrifugation and the 18S and 14S forms were collected together. On standing the 11S form began to appear with a decrease in 18S and 14S forms. Thus the 18S and 14S forms even in a purified state can convert to the 11S form. This does not rule out a proteolytic conversion, (4,5,6) because our procedure could have trapped, released and reactivated serine proteases (Fig. 3). The 18S and 14S forms appear to have globular heads attached to a tail as judged by electron microscopy (10). The 11S form is a globular tetramer.

PURIFICATION OF ACETYLCHOLINESTERASE

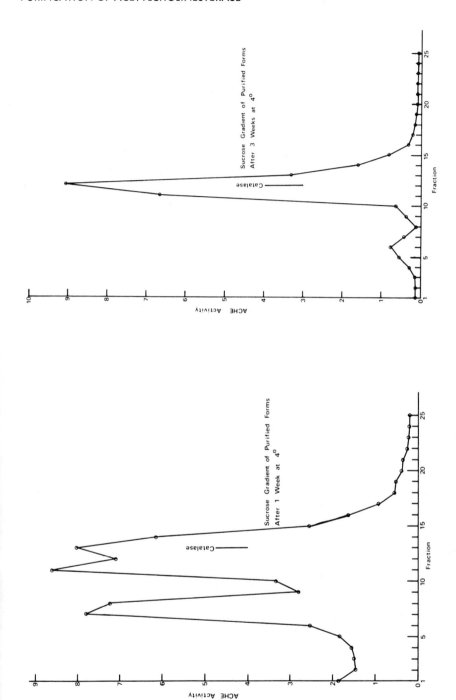

Figure 3. Sucrose gradient profiles of purified 18S and 14S forms after standing at 4° for the indicated times.

In another experiment the 18S form was separated and subject to "active enzyme" sedimentation using an analytical ultracentrifuge (Spinco Model E) (7,8,9). In this procedure a lamella of enzyme solution is layered on top of solution containing a chromogenic substrate and the zonal sedimentation of enzyme is monitored by the optical absorbance of the product of enzyme activity. The Ellman method for assaying acetylcholinesterase is equivalent to a method with a chromogenic substrate and was used in this procedure. We were thus able to get a rather precise sedimentation coefficient for the 18S form. We obtained 18.3S. We were able to answer a question of some significance concerning the aggregated form of this enzyme. Although they are stable in high salt the 18S and 14S forms aggregate reversibly in low or moderate salt < 0.3 M (4,5,6). Is this aggregated form active? Using active enzyme sedimentation we are able to show that the aggregated form has a sedimentation coefficient of 70.S and is precisely as active as in its non aggregated forms. This observation is of some significance because it indicates that the catalytic units in the aggregate are not randomly joined but all present their active sites to the solution phase where they are available to substrate. The aggregate is an organized particle. There are, of course, degrees of organization. Perhaps the simplest organization is one in which the aggregate resembles a spherical micelle with relatively hydrophobic tails in the interior and relative hydrophilic catalytic units on the outside. Another simple possibility is that the particle is cylindrical with the tails forming a relatively hydrophobic core. We have suggested that hydrophobicity may be the driving force for aggregation but, in fact, there is little evidence for this because the effect of salt would not be explained and if anything would appear to be in the opposite direction. However if we assume that hydrophobicity is almost sufficient to produce aggregation and allow for the formation of ionic bonds in low salt we can account for the aggregation. Of course there may be a higher order or organization based upon more specific interactions and such a possibility is of especial interest in connection with the question of the native form of the enzyme as part of an active membrane.

We have measured the K_m of the 18S, 14S and 11S forms. All have the same value, 1×10^{-4} M in 0.25 M ionic strength and 1.4×10^{-4} M in 1.0 M ionic strength pH 7.0.

References

1. P. Cuatrecasas, Adv. in Enzymology 36, 29-89 (1972).
2. Y. Ashani and I. B. Wilson, Biochim. Biophys. Acta 276, 317 (1972).
3. H. F. Voss and Y. Ashani, Fed. Proc. 32, 553 Abs. (1973).

4. J. Massoulié and F. Rieger, Eur. J. Biochem. 11, 441 (1969).
5. F. Rieger, S. Bon and J. Massoulié, C. R. Acad. Sci. Paris 274D, 1753 (1972).
6. F. Rieger, S. Tsuji and J. Massoulié, Eur. J. Biochem. 14, 430 (1970).
7. J. Vinograd, R. Radloff and R. Bruner, Biopolymers 3, 481 (1965).
8. R. Cohen, B. Girand and A. Messiah, Biopolymers 5, 203 (1967).
9. S. L. Snyder, I. B. Wilson and W. Bauer, Biochim. Biophys. Acta 258, 178 (1972).
10. F. Rieger, S. Bon and J. Massoulié, Eur. J. Biochem. 34, 539 (1973).

COOPERATIVE EFFECTS OF AMP, ATP AND FRUCTOSE 1,6-P_2 ON THE SPECIFIC

ELUTION OF FRUCTOSE 1,6-DIPHOSPHATASE FROM CELLULOSE PHOSPHATE

Joseph Mendicino and Hussein Abou-Issa

Department of Biochemistry, University of Georgia

Athens, Georgia 30602

In 1963 Mendicino and Vasarhely (1) first observed that fructose 1,6-diphosphatases from kidney and liver were specifically bound to cellulose phosphate and carboxymethylcellulose. These studies further clearly demonstrated that AMP and high concentrations of fructose 1,6-P_2 inhibited the activity of the enzyme (1). Since this time a number of other workers have utilized the specific binding of this enzyme to anion exchange columns as the principal purification step in the isolation of the enzyme from a large number of different tissues (2,3,4,5).

In more recent experiments we observed that incubation of rat liver mitochondria with ATP and Mg^{++} resulted in the formation of products which were capable of eluting fructose 1,6-diphosphatase from cellulose phosphate columns (6,7). The binding of very small amounts of ATP and AMP which was formed from ATP by the action of myokinase and ATPase which are associated with the mitochondria caused fructose 1,6-diphosphatase to be released from the column. The elution of the enzyme from the phosphoryl groups in the column occurred because of a cooperative effect between ATP and AMP. The amount of enzyme released from the column in the presence of very small amounts of both nucleotides was much greater than the amount eluted by higher concentrations of each nucleotide alone.

This surprizing and unexpected finding of specific cooperative effects upon the binding properties of fructose 1,6-diphosphatase was examined. These kinds of effects are usually associated with alterations in the kinetic properties of enzymes. Evidently, the binding properties of specific groups on enzyme surfaces can also be changed by cooperative interactions between substrates and allosteric effectors.

Influence of NaCl, MgCl₂ and AMP on the Elution Profiles of Fructose 1,6-Diphosphatase. The data obtained in previous studies suggested that the substances responsible for the elution of fructose 1,6-diphosphatase from cellulose-phosphate were adenine nucleotides. The results of experiments designed to compare the relative abilities of AMP and the chloride salts of a monovalent cation, Na⁺, and a divalent cation, Mg⁺, to elute fructose 1,6-diphosphatase from cellulose phosphate are summarized in Fig. 1.

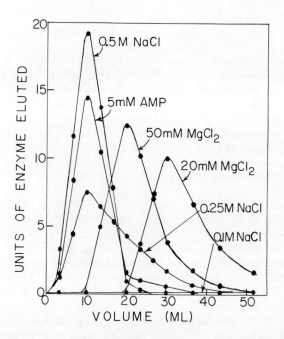

Figure 1. Elution of D-fructose 1,6-diphosphatase from cellulose phosphate columns with varying concentrations of NaCl, MgCl₂ and AMP. About 19 units of D-fructose 1,6-diphosphatase was applied to several cellulose phosphate columns (2.0 x 4.5 cm). The columns were washed with 20 ml of 0.05 M Tris-HCl, pH 8.0 and the enzyme was eluted, in each case, with the indicated concentration of each compound dissolved in this same buffer. Fractions of 5 ml were collected at a flow rate of 2 ml per min and they were assayed for enzyme activity by the standard procedure.

NaCl was a very effective eluant at 0.5 M and it removed virtually all of the enzyme from the cellulose phosphate column in a small volume. It was much less effective at 0.25 M and considerable tailing was observed at this concentration. No enzyme was eluted with 0.1 M NaCl. Fructose 1,6-diphosphatase presumably binds to cellulose phosphate because cationic groups on the surface of the enzyme react with the anionic phosphoryl groups in the column. Salts, such as NaCl, displace the enzyme by competition for the anionic phosphoryl binding sites in the cellulose phosphate. AMP, at 5 mM was nearly as effective as 0.5 M NaCl in eluting the enzyme under identical conditions. The mechanism in this case must involve a specific interaction of this allosteric inhibitor with the enzyme rather than a competition with cationic sites on the enzyme for phosphoryl groups immobilized in the column. $MgCl_2$ at a concentration of 50 mM, also eluted the enzyme from the column. However, the peak containing the enzyme emerged somewhat later than those eluted with AMP or NaCl. Magnesium ion is required for enzymatic activity, and this divalent metal ion probably binds to both the enzyme and the phosphoryl groups in the column. At 20 mM $MgCl_2$ the peak containing enzyme activity was retarded still further. This observation may be explained if the amount of magnesium ion which binds to the enzyme is the same at both concentrations of $MgCl_2$, while the amount which competes with the enzyme for the phosphoryl groups of the column is greater at 50 mM. Thus, the number of phosphoryl groups which would be free to react with the enzyme would be less at 50 mM $MgCl_2$. The magnesium-enzyme complex may bind more tightly to phosphoryl groups in the column than the free enzyme and this would result in some retardation of the enzyme in the column whenever magnesium ion was present in the eluting solution.

Influence of the Concentration of ATP on the Binding of Fructose 1,6-Diphosphatase. The cooperative effects of ATP and AMP on the binding of the enzyme were examined. The influence of the concentration of ATP, alone and in the presence of AMP on the binding of fructose 1,6-diphosphatase to cellulose phosphate is shown in Fig. 2. The amount of fructose 1,6-diphosphatase eluted, in each case, was determined by a standardized assay method. Stored preparations of fructose 1,6-diphosphatase were routinely activated by incubation for 15 min with 33 mM Tris-HCl, pH 8.0 and 67 mM cysteine, pH 8.0 at room temperature just before they were used in these experiments. This procedure insured an accurate measure of the total amount of enzyme which was bound to the column. About 18 to 20 units of the activated enzyme was diluted to 20 ml with 0.05 M Tris-HCl, pH 8.0 and it was passed into a small cellulose phosphate column (2.0 x 4.5 cm), and the column was washed with 20 ml of 0.05 M Tris-HCl, pH 8.0. Under these conditions all of the fructose 1,6-diphosphatase activity was adsorbed to the column.

The compounds to be tested for their ability to promote the

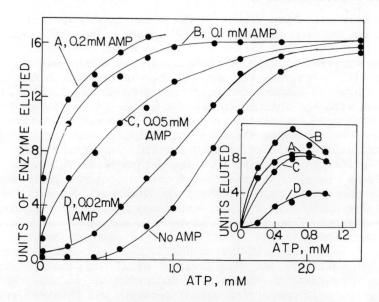

Figure 2. Influence of the concentration of ATP on the binding of D-fructose 1,6-diphosphatase to the phosphoryl groups of cellulose phosphate in the presence of 0.02 mM, 0.05 mM, 0.1 mM and 0.2 mM AMP. The difference between the sum of the activities eluted with each nucleotide alone and that eluted with both compounds at each concentration is shown in the inset. About 16 units of D-fructose 1,6-diphosphatase were applied to each column.

dissociation of the enzyme from the phosphoryl groups in the column were dissolved in 30 ml of cold 0.05 M Tris-HCl, pH 8.0. This solution was passed through the column, and the column was washed with 20 ml of 0.05 M Tris-HCl, pH 8.0, to remove all of the dissociated enzyme. The eluting solution and wash were combined and assayed for fructose 1,6-diphosphatase activity. The fructose 1,6-diphosphatase which remained bound to the column was then removed with 30 ml of 0.05 M Tris-HCl, pH 8.0 - 0.5 M NaCl. The total activity eluted in these samples was always equal to the amount of enzyme applied to the column.

As seen in Fig. 2, ATP alone eluted a small amount of the enzyme at 0.6 mM, and the amount of the enzyme eluted increased with increasing concentrations of ATP. At 2 mM ATP nearly all of the enzyme was eluted under these conditions. The addition of AMP greatly facilitated the elution of the enzyme, and much more fructose 1,6-diphosphatase was eluted at low concentrations of ATP. At 0.6 mM ATP little or no enzyme was eluted and AMP alone eluted 0.5 units of enzyme at 0.02 mM, 1.5 units at 0.05 mM, and 3.0 units at 0.10 mM. However, 0.6 mM ATP and 0.02 mM AMP together eluted 4.0 units of enzyme, with 0.05 mM AMP, 10.5 units were eluted and with 0.01 mM AMP, 13.6 units were eluted. Apparently, the binding of both AMP and ATP to the enzyme caused a conformational change which decreased its affinity for phosphoryl groups much more then either one alone. The results summarized in the inset in Fig. 2 show the difference between the amount of fructose 1,6-diphosphatase eluted with various combinations of ATP and AMP after the amounts eluted by each one alone have been subtracted. At high concentrations the cooperative effect became much less apparent. At 0.1 mM AMP and 0.6 mM ATP the difference in the amount

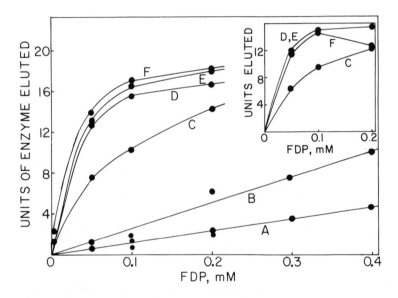

Figure 3. Influence of the concentration of D-fructose 1,6-P_2 on the binding of D-fructose 1,6-diphosphatase to the phosphoryl groups of cellulose phosphate in the presence of: A, no ATP or AMP; B, 0.2 mM ATP; C, 5 μM AMP; D, 10 μM AMP; E, 20 μM AMP; F, 50 μM AMP. The difference between the sum of the activities eluted with each effector alone and that eluted with both compounds at each concentration is shown in the inset. A total of 18 units of D-fructose 1,6-diphosphatase was adsorbed to each column.

of fructose 1,6-diphosphatase eluted was 11 units. When the AMP concentration was increased to 0.2 mM, the difference in the amount of fructose 1,6-diphosphatase eluted decreased to 8 units as seen in the inset. The cooperative effects between AMP and ATP could be best observed at low concentrations where neither nucleotide alone eluted very much enzyme.

Influence of the Concentration of Fructose 1,6-P_2 on the Binding of Fructose 1,6-Diphosphatase. The effect of increasing concentrations of fructose 1,6-P_2 alone and in the presence of 5 µM, 10 µM, 20 µM and 50 µM AMP on the binding of fructose 1,6-diphosphatase is shown in Fig. 3. Fructose 1,6-P_2 alone, curve A, eluted about as much enzyme as an equivalent concentration of AMP, and it was much more effective than ATP alone. The addition of 2 mM ATP, curve B, enhanced elution by fructose 1,6-P_2 only two-fold. However, AMP at much lower concentrations, greatly increased the amount of enzyme eluted by fructose 1,6-P_2. Only 2.3 units of enzyme were eluted with 0.2 mM fructose 1,6-P_2. About 5.1 units were eluted with 0.2 mM fructose 1,6-P_2 and 0.2 mM ATP. At the same concentration of fructose 1,6-P_2, the addition of only 5 µM AMP resulted in the elution of 14 units of enzyme, while 10 µM AMP and 2 mM fructose 1,6-P_2 eluted 16.6 units. At higher concentrations of AMP, 20 µM and 50 µM, nearly all of the enzyme was eluted from the column. The data in the inset in Fig. 3 show the difference between the amount of enzyme eluted with various combinations of AMP and fructose 1,6-P_2 after the amount of enzyme eluted by each compound alone is subtracted. Again, it may be seen that the cooperative effects are best observed at very low concentrations of AMP and fructose 1,6-P_2.

Effect of Increasing Concentrations of AMP on the Binding of Fructose 1,6-Diphosphatase. The influence of very low concentrations of AMP on the binding of fructose 1,6-diphosphatase to phosphoryl groups in cellulose phosphate is shown in Fig. 4 and Fig. 5. Very little enzyme was eluted at concentrations of AMP below 20 µM. However, at higher concentrations AMP alone was about as effective as fructose 1,6-P_2 in eluting the enzyme and it was much more effective than ATP. About 60% of the enzyme, 10 units, was eluted with 0.4 mM AMP, whereas 0.4 mM ATP did not elute any enzyme from the column. The addition of relatively high concentrations of ATP enhanced elution by AMP, as shown in Fig. 4. In the presence of 0.2 mM ATP and 0.1 mM AMP approximately 9 units of enzyme were eluted. As seen in Fig. 5 only 25 µM fructose 1,6-P_2 and 10 µM AMP were required to elute about the same amount of enzyme. At this concentration neither AMP nor fructose 1,6-P_2 alone will elute the enzyme. These results clearly show that combinations of ATP, AMP and the substrate, fructose 1,6-P_2, are much more effective in altering the binding of fructose 1,6-diphosphatase to cellulose phosphate than any one of them alone. At physiological concentrations of ATP, 1-2 mM, and AMP, 0.05 - 0.1 mM variations in the

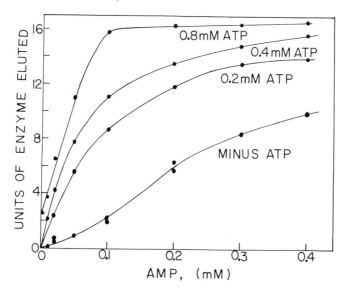

Figure 4. Influence of low concentrations of AMP on the binding of D-fructose 1,6-diphosphatase to the phosphoryl groups of cellulose phosphate in the presence of 0.2 mM, 0.4 mM, and 0.8 mM ATP. About 16 units of D-fructose 1,6-diphosphatase were applied to each column.

concentration of fructose 1,6-P_2 between 10 µM and 50 µM could regulate the binding of this enzyme to free phosphoryl groups.

Specificity for the Binding of Fructose 1,6-Diphosphatase to Phosphoryl Groups. The effects observed on the binding of fructose 1,6-diphosphase to cellulose phosphate were specific for AMP and ATP. Other nucleotides and ADP either alone or in various combinations were all ineffective in promoting the dissociation of fructose 1,6-diphosphatase from cellulose phosphate columns. As expected, the elution effects were also specific for fructose 1,6-P_2. Glucose 6-P, Pi, P-Pi, fructose 6-P, glucose 1-P, P-enolpyruvate and 3-P-glycerate were all inactive at a concentration of 1 mM.

The addition of 0.8 mM $MgCl_2$ had little or no effect on the amount of enzyme eluted by AMP, ATP and fructose 1,6-P_2 either alone or in various combinations.

The binding of these metabolites to kidney fructose 1,6-

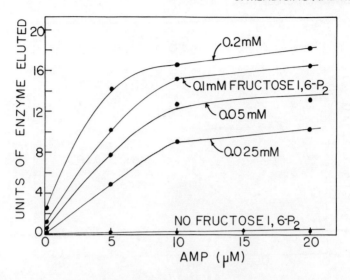

Figure 5. Influence of low concentrations of AMP on the binding of D-fructose 1,6-diphosphatase to the phosphoryl groups of cellulose phosphate in the presence of 0.025 mM, 0.05 mM, 0.10 mM and 0.20 mM fructose 1,6-P_2. The standard assay was used and 16 units of D-fructose 1,6-diphosphatase were applied to the columns.

diphosphatase was also examined by the gel filtration technique. The enzyme contained four binding sites per mole for each compound. The K_D for AMP was 9.7 µM, and in the presence of 0.2 mM fructose 1,6-P_2 it was decreased to 5.2 µM. The K_D for AMP increased to 18.5 µM in the presence of 0.4 mM ATP. The addition of fructose 1,6-P_2 and ATP did not affect the number of AMP molecules bound, which suggested that the binding sites for these compounds might be adjacent but not identical. The K_D for fructose 1,6-P_2 was 8.5 µM, and it decreased to 3.6 µM in the presence of 0.1 mM AMP. It decreased to 7.0 µM in the presence of 0.4 mM ATP. ATP which was the least effective in altering the binding of fructose 1,6-diphosphatase to cellulose phosphate and a K_D of 18.3 µM was obtained for this compound. In the presence of 0.2 mM fructose 1,6-P_2 or 0.2 mM AMP the K_D for ATP increased to 43.7 µM. The binding properties of these ligands to the enzyme generally correlated with their ability to decrease the binding of fructose 1,6-diphosphatase to phosphoryl groups in cellulose phosphate columns.

In order to correlate the binding of AMP and fructose 1,6-P_2 with the inhibition of catalytic activity by AMP kinetic studies were carried out. The Km of the enzyme for fructose 1,6-P_2 was found to be 6.1 μM in the absence of AMP. In the presence of 0.01 mM AMP, the Km for fructose 1,6-P was 5.7 μM and with 0.02 M AMP it decreased to 4.1 μM. These differences were not quite as large as those observed in the binding studies. The K_D of the enzyme for fructose 1,6-P_2 was 8.5 μM and it decreased to 3.6 μM in the presence of 0.1 mM AMP. However, these results further support the observation that AMP increases the affinity of fructose 1,6-diphosphatase for fructose 1,6-P_2, even though AMP is also a potent inhibitor of the enzyme.

DISCUSSION

From the results obtained in these studies it is possible to conclude that fructose 1,6-diphosphatase does bind very tightly and specifically to the phosphoryl groups of cellulose phosphate columns. The binding of this enzyme can be preferentially influenced by either its substrate or an allosteric inhibitor, AMP, and large cooperative effects are observed when very low concentrations of both compounds are present. The results also clearly show that the enzyme contains at least two or three different binding sites which do interact with each other. The native enzyme contains 4 identical subunits (8) and the present study shows that a maximum of 4 equivalents each of AMP, ATP and fructose 1,6-P_2 can bind to the enzyme. The results further indicate, but do not prove, that enzyme-substrate, enzyme-AMP and AMP-enzyme-substrate complexes are unable to bind to phosphoryl groups in cellulose phosphate columns.

One of the problems in understanding the interaction between the surface of the enzyme and modified cationic cellulose columns is the difficulty in estimating the number of sites on the enzyme which are displacing appropriate ions in the column. What is more, the actual dimensions of the surfaces which are interacting cannot be established with any certainty. In the case of fructose 1,6-diphosphatase very high salt concentrations greater than 0.2 M NaCl, are required to elute the enzyme from cellulose phosphate columns. One would imagine, as a first approximation, that a large number of charged groups on the enzyme were interacting with cationic groups in the modified cellulose column. However, a very few high affinity binding sites could also explain the results obtained in the present study.

The question arises as to whether each of the binding sites for AMP and fructose 1,6-P_2 can bind to phosphoryl groups or whether only one interacts while the other and ATP are effective only because they are able to convert the enzyme to a form which decreases the affinity of the bound site for the phosphoryl group. Several recent observations indicate that the latter explanation may be correct. Treatment of fructose 1,6-diphosphatase with pyridoxal 5'-phosphate and subsequent reduction with sodium borohydride in the presence of fructose 1,6-P_2 has been shown to result in the formation of an enzymatically active derivative of the enzyme which contains about 4 moles of pyridoxal-5'-phosphate per mole of enzyme. The catalytic properties of the enzyme are not altered by this treatment, however its sensitivity to allosteric inhibition by AMP is greatly decreased (9). These studies further showed the modified enzyme no longer bound AMP, while the binding of fructose 1,6-P_2 to the enzyme was unaffected by this modification. In the present studies we have noticed that the binding of this modified enzyme to cellulose phosphate was also greatly decreased, which indicates that the 4 highly reactive ε-aminolysyl residues which react with pyridoxal-5'-phosphate (9) may also be required for binding of the enzyme to phosphoryl groups in the column.

Inhibition by AMP and binding of the enzyme to cellulose phosphate were both effected by this treatment, whereas the substrate site and enzymatic activity were not.

Another observation made in the present studies also indicates that only the AMP site may be bound to phosphoryl groups in the column. Fructose 1,6-diphosphatase bound to cellulose phosphate was found to be enzymatically active when it was assayed with a coupled assay which contains a fructose 1,6-P_2 regenerating system (1). It had almost the same kinetic properties as the free enzyme under these conditions. These observations are consistent with the proposal that only cationic groups in the AMP binding site may be reacting with phosphoryl groups in the column.

The possible physiological significance of the observations made in the present study on the regulation of carbohydrate metabolism may be considered. It has been suggested that the activity of fructose 1,6-diphosphatase might be regulated by changes in the concentration of AMP or fructose 1,6-P_2 during gluconeogenesis. However, evidence obtained on the level of these metabolites in liver and kidney and the kinetic properties of the enzyme indicate that an inhibition by AMP would exist most of the time and that the highest levels of fructose 1,6-P_2 found in these tissues 0.03 µmole per g wet weight (10) would be too low to influence the activity of the enzyme. In other studies it was proposed that changes in the sensitivity of the enzyme to inhibition by AMP, rather than changes in the concentration of AMP might regulate the activity of the enzyme. However, here again the concentrations of metabolites in these tissues did not correspond to the amounts required to effect the activity of the isolated enzyme. The role of AMP and fructose 1,6-P_2 in the physiological regulation of fructose 1,6-diphosphatase remains obscure.

The binding of fructose 1,6-diphosphatase to phosphoryl groups on cellulose phosphate columns and the specific effects of AMP and fructose 1,6-P_2 on the binding may be a clue as to a possible means of regulating the activity of the enzyme in vivo. The results of some preliminary experiments indicate that the regulation of fructose 1,6-diphosphatase and P-fructokinase involve an interaction between these two proteins and other proteins or subcellular particles present in swine kidney homogenates. The concentrations of fructose 1,6-P_2, AMP, and ATP required to alter the binding of fructose 1,6-diphosphatase to cellulose phosphate, were all within the observed physiological range of concentrations observed in kidney. The intracellular concentration of FDP varies between 5 to 26 µM (10). The concentrations of AMP and ATP apparently do not change in vivo (10). Perhaps at low concentrations of fructose 1,6-P_2 the enzyme would be bound to other proteins or membranes by its AMP binding site and it would then be active and insensitive to inhibition by AMP. At somewhat higher concentrations of fructose 1,6-P_2 which increases during gluconeogenesis, the enzyme would be released and AMP would then inhibit the enzyme under these conditions. Thus, the binding or release of fructose 1,6-diphosphatase which may be regulated by the level of fructose 1,6-P_2 would control the activity of the enzyme.

Fructose 1,6-diphosphatase does not bind very tightly to other proteins in the cell and a complex has not been isolated as such. Loose complexes would probably be ruptured during the preparation of homogenates and separated during purification. The allosteric areas of the enzyme may be exposed under these conditions and make it appear that AMP is the regulator when the actual regulation may be more complicated. The loss of protein interactions which usually involve the surfaces of enzymes may even alter the allosteric properties of the enzyme while having no effect on its catalytic properties. The allosteric properties are dependent on the nature and conformation of the exposed enzyme surface which is more likely to be altered by compounds in reaction mixtures or by isolation procedures. The active site, on the other hand, is usually more rigid and somewhat less exposed. In crude swine kidney extracts, fructose 1,6-diphosphatase is relatively stable and there is no requirement for sulfhydryl reagents for maximum activity. Perhaps this could be explained by the removal of a protein during purification which is normally attached to the enzyme in the cell. Sulfhydryl reagents may satisfy this requirement of binding to other proteins.

It may be postulated that the sites on the enzyme which bind to phosphoryl groups in cellulose phosphate, may be the same as those which bind to the effectors and substrate. These may also be sites which bind to proteins or membranes in the cell. Enzymes which are related by function do bind tightly together and can be isolated as complexes, as for example, pyruvate oxidase, fatty acid synthetase, and acetyl CoA carboxylase. Perhaps many other enzymes, related to each other by function in the cell, are also bound in complexes but much more loosely so that they are not isolated as such. In the tightly bound complexes, binding appears to have little effect on activity but probably facilitates the function by maintaining the enzymes in close proximity. Perhaps in loose complexes, where binding would be more easily reversed, the binding itself may assist in regulating the activity of the enzyme.

SUMMARY

The effect of AMP, ATP and fructose $1,6$-P_2 on the binding of swine kidney fructose 1,6-diphosphatase to phosphoryl groups in cellulose-phosphate columns was studied with a standardized elution technique. All three compounds interacted with the enzyme specifically and decreased its binding to cellulose-phosphate columns. However, AMP and fructose $1,6$-P_2 exhibited the highest activity in decreasing the affinity of the enzyme for phosphoryl groups. These effects were specific for AMP, ATP and fructose $1,6$-P_2 and other metabolites including various nucleotides, fructose 6-P, glucose 6-P and ADP did not significantly influence the binding properties of the enzyme. The effects were cooperative, since combinations of any two of these compounds decreased the binding of fructose 1,6-diphosphatase much more than the sum of each one alone. These properties may result from the formation of substrate-enzyme-allosteric effector complexes which cause a change in the conformation of the enzyme and decrease the charge of specific cationic groups on amino acid residues in the AMP and fructose $1,6$-P_2 binding sites. These effects on the binding properties of fructose 1,6-diphosphatase were observed with very low concentrations of AMP, 20 µM, and fructose $1,6$-P_2, 20 µM. It has been calculated that at physiological concentrations of ATP and AMP a variation of fructose $1,6$-P_2 concentration between 10 µM and 50 µM could regulate the release or binding of fructose 1,6-diphosphatase to free phosphoryl groups.

FOOTNOTE: The following abbreviations were used: AMP, adenosine monophosphate; ATP, adenosine triphosphate; fructose $1,6$-P_2, fructose 1,6-diphosphate.

REFERENCES

1. Mendicino, J. and Vasarhely, F. (1963) J. Biol. Chem., 240, 3528.
2. Black, W. J., Van Tol, A., Fernando, J. and Horecker, B. L. (1972) Arch. Biochem. Biophys., 151, 576-590.
3. Marcus, F. (1967) Arch. Biochem. Biophys., 122, 393-399.
4. Traniello, S., Pontremoli, S., Tashima, Y. and Horecker, B. L. (1971) Arch. Biochem. Biophys., 146, 161-166.
5. Sia, C. L. and Horecker, B. L. (1972) Arch. Biochem. Biophys., 149, 222-231.

6. Mendicino, J., Prihar, H. S. and Salama, F. M. (1968) J. Biol. Chem., 243, 2710-2717.
7. Kratowich, N. and Mendicino, J. (1970) J. Biol. Chem., 245, 2483-2492.
8. Mendicino, J., Kratowich, N. and Oliver, R. M. (1972) J. Biol. Chem., 247, 6643-6650.
9. Giovanna, C., Hubert, E. and Marcus, E. (1972) Biochemistry, 11, 1796-1801.
10. Exton, J. H. and Park, C. R. (1969) J. Biol. Chem., 244, 1424-1431.

AN ANALYSIS OF AFFINITY CHROMATOGRAPHY USING IMMOBILISED ALKYL NUCLEOTIDES

P. D. G. Dean, D. B. Craven, M. J. Harvey & C. R. Lowe

Dept. of Biochemistry, The University

Liverpool, England L69 3BX

Group specific matrices for affinity chromatography have now received considerable attention (1-9). Some of these studies have provided amongst the best examples of this aspect of affinity chromatography (10). Amongst the group specific matrices and in particular amongst the immobilised cofactors, the nicotinamide nucleotides present a very wide choice of applications and of different enzyme systems to study. However, because of the chemical complexity of the nicotinamide nucleotides, severe restrictions are imposed on the potential methods available for their immobilisation. These nucleotides may be immobilised readily onto cellulose or Sephadex or agarose containing aminocaproic acid (1,11,12). The resultant polymers are almost certainly of undefined nature (13) apart from being contaminated with unreacted carboxyl groups. The former need not necessarily be a drawback; the polymer preparation described by Allen & Majeries (10) does not specify which carboxyl groups of B_{12} are utilised in the linkage to the agarose and this undefined matrix is capable of purification factors of thousands. This is not so with nucleotides where the purification factors rarely exceed fifty times (1,8).

Abbreviations:
LDH - lactate dehydrogenase (H_4 - pig heart, M_4 - rabbit muscle); YADH - yeast alcohol dehydrogenase; G6PDH - D-glucose 6-phosphate dehydrogenase; MDH - L-malate dehydrogenase; G3PDH - D-glyceraldehyde 3-phosphate dehydrogenase; GDH - L-glutamate dehydrogenase; GR - glutathione reductase; GK - glycerokinase; MK - myokinase; HK - hexokinase; BSA - bovine serum albumin; N^6-AMP - N^6-(6-aminohexyl) 5'-AMP; P^2-ADP - P^1-(6-aminohexyl)-P^2-(5'-adenosine)-pyrophosphate.

From early studies in our laboratory comparing a defined AMP polymer, kindly donated by Klaus Mosbach, with undefined ones we concluded that the potential of the former far exceeded the undefined matrices both in interpretation of binding characteristics and applications to enzymology (Fig. 1) (4). What was particularly exciting to us was the possibility of using these systems as probes of enzyme structure. This remains a relatively unexplored field.

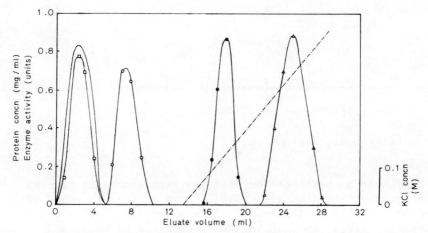

Fig. 1. Chromatography of crude yeast extract on N^6-AMP-Sepharose. (□) G6PDH, (△) ADH, (○) GR, (●) MDH, (——) Protein & (—·—) KCl gradient.

Thus a sounder approach to the immobilisation of nucleotides is to prepare derivatives which are capable of specific immobilisation; for example, ones containing reactive alkyl amines. Several research groups have now reported such matrices (Table 1).

Later in our work we needed a supply of the aminoalkyl-AMP analogues and set out to devise a process from readily available chemicals using as few steps as possible. Originally we had in mind a reaction involving the displacement of SO_3H or such a leaving group at C^6 with an alkyl amine to give the corresponding N^6-alkyl derivatives. However, we found that the starting material reacted without the need to increase its potential as a leaving group. The starting material was thioinosine riboside 5'-monophosphate (I) in aqueous diamine (Fig. 2).

Fig. 2

Table 1

Ligand	Support	Linkage	Authors	Date	Ref.
NAD	Cellulose	Diazo	Lowe & Dean	1971	1
NAD	Sepharose	Ribose	Dean & Lowe	1971	13
NAD	Glass	Diazo	Weibel, Weetall & Bright	1971	14
NAD	Sepharose	Diazo	Hocking & Harris	1972	15
AMP	"	N^6-Alkyl	Guildford, Larsson & Mosbach	1972	16
ADP	"	P^2-Alkyl	Barker, Olsen, Shaper & Hill	1972	17
ATP	"	P^3-O-Phenyl	Berglund & Eckstein	1972	18
NADP	"	Ribose	Lamed, Levin & Wilchek	1973	19
AMP	"	N^6-Alkyl	Craven, Harvey, Lowe & Dean	1973	20
NAD	"	N^6-Alkyl	Craven, Harvey & Dean	1973	21

The yield of aminoalkyl-AMP after 16hr in a sealed ampoule at $90°C$ was 80%. The reaction has interesting properties in that it is surface dependent (scale-up requires the addition of glass wool to the reaction vessel). By varying the number of methylene groups (n) in the diamine we can prepare a range of AMP derivatives, the nucleotide of which is placed further and further from the point of immobilisation; thus we have prepared the series C_2 to C_{12} aminoalkyl-AMPs (22).

This synthesis has led on to the preparation of other nucleotides, ATP, ADP and NAD^+. The latter is most interesting to us and we have used a method based on trifluoracetic anhydride as a catalyst:

AMINOALKYL-AMP + NMN + TFA ⟶ AMINOALKYL-NAD
       or
       NND   25% yield
       or
       NAD

Conditions: Anhydrous, room temperature, sealed ampoule, 16 hr.
Isolation of product: Dowex-1 to remove AMP; electrophoresis separates aminoalkyl-NAD from remaining contaminants.

The use of NAD in this synthesis is only an economic one; it poses considerable problems in the separation of products. The aminoalkyl-NAD has been characterised by U.V. spectra in the oxidised and reduced states, and in the presence of cyanide and sulphite (see Fig. 3). The R_f values of aminoalkyl-NAD in eight T.L.C. systems have been compared with NAD, NMN, NND and Aminoalkyl AMP (21). The alkyl NAD is enzymically reducible with a variety of enzyme systems; the best involving yeast alcohol dehydrogenase and the derivatised NAD is readily immobilised in high concentrations onto cyanogen bromide activated agarose.

Some of the most interesting properties of aminohexyl-AMP and aminohexyl-NAD, apart from their use in immobilised reactors, include their potential binding characteristics. We have always assessed these properties by using salt or nucleotide gradients

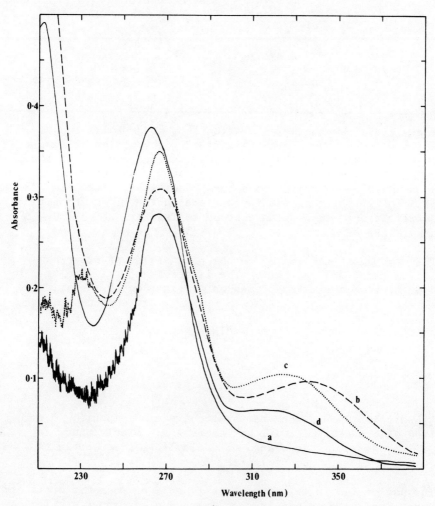

Fig. 3. Absorption spectra of N^6-(6-aminohexyl)-NAD^+. (a) Oxidised (b) Reduced (c) Cyanide adduct (d) Sulphite adduct.

in an attempt to quantitate the interaction of enzymes with individual matrices. The definition of β is the binding expressed as an eluant concentration at which the peak of enzyme activity is eluted. Initially it was decided to examine the AMP matrices and we made attempts primarily to examine the importance of the position of attachment to the polymer matrix.

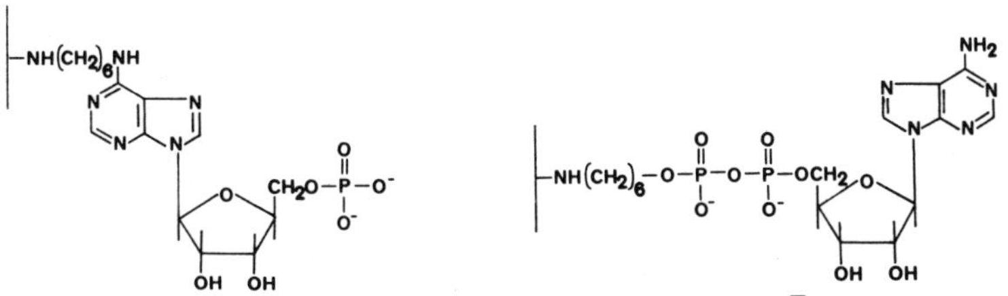

Fig. 4. A comparison of the structure of (I) N^6-(6-aminohexyl)-5'-AMP and (II) P^1-(6-aminohexyl)-P^2-(5'-adenosine)-pyrophosphate.

Table 2. Enzyme binding to two defined immobilised nucleotide matrices.
[I] N^6-(6-aminohexyl) 5'-AMP-Sepharose (1.5μmoles AMP/ml)
[II] P^1-(6-aminohexyl)-P^2-(5-adenosine)-pyrophosphate-Sepharose (6.0μmoles AMP/ml).

ENZYME	SOURCE	BINDING (β) I	BINDING (β) II
Lactate dehydrogenase	Pig heart	1M(a)	1M(a)
Lactate dehydrogenase	Rabbit muscle	1M(a)	1M(a)
D-glucose 6-P dehydrogenase	Yeast	0.	170
Malate dehydrogenase	Pig heart	65	490
Alcohol dehydrogenase	Yeast	400	0
D-glyceraldehyde 3-P dehydrogenase	Rabbit muscle	0	1M(a)
3-Phosphoglycerate kinase	Yeast	70	260
Pyruvate kinase	Rabbit muscle	100	110
Hexokinase	Yeast	0	0
Creatine kinase	Rabbit muscle	0	0
Myokinase	Rabbit muscle	0	380
Glycerokinase	Candida mycoderma	122	0

(a) Elution effected by a 200μl pulse of 5×10^{-3}M NADH.

Table 2 compares the binding of two adenine nucleotide derivatives (Fig. 4) prepared by N^6-immobilisation as described above and via P^2-immobilisation as described by Barker, Olsen, Shaper & Hill (17).

Notable is the behaviour of lactate dehydrogenase (LDH) where very tight binding is observed for both polymers. The binding to N^6-AMP has led to an interesting purification procedure for this enzyme (6, q.v. 23). Since the binding could not be measured by means of a salt gradient up to 1 Molar, we have sought other ways of measuring the interaction of LDH for these matrices.

Dilution of the polymers provides some clue: with the N^6-alkyl-AMP a lower value for β is obtained on dilution. However, if the gel is diluted with unsubstituted beads 'β' does not fall as far and indeed reaches a plateau (the height of which is related to the original ligand concentration) however when the dilution is carried out by placing less and less ligand uniformly throughout the gel then 'β' falls to zero at a concentration around 16 mµmoles /ml (Fig. 5).

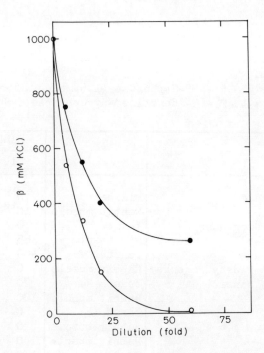

Fig. 5. Effect of dilution on the binding of LDH-M_4 to N^6-AMP-Sepharose.
(o) Ligand uniformly coupled (●) Ligand diluted with Sepharose

The stronger binding observed with the former polymers probably reflects the manner of dilution since the pockets of ligand still contain concentrations of ligand equal to that of the undiluted gel. These pockets according to the kinetics of reversible equilibria, would maintain the enzyme in an immobilised form by restricting its dissociation from the matrix. When compared with the P^2-immobilized nucleotide in this way the N^6-alkyl matrix binds M-4 LDH more tightly indicating the preference for an uncluttered phosphate when binding AMP analogues. Indeed, inspection of the LDH 2Å model indicates that, assuming AMP analogues bind in the same way as the AMP portion of NAD, enough space is available for the extension arm to get outside the LDH protein despite the two very different positions of attachment (Rossmann & Dean, unpublished). We have examined this effect with LDH further and showed a sigmoidal response in the plot of capacity of the matrix versus concentration of uniformly coupled ligand (24) which suggests a cooperativity phenomenon in the binding process. However, the dilution experiment used to differentiate between the two polymers was carried out with unsubstituted beads. Under these conditions the plot of capacity versus ligand concentration is not sigmoidal which indicates a qualitatively different interaction and supports our previous conclusion that dilution with unsubstituted beads leads to isolated pockets of ligand which behave as if they were undiluted; albeit the parameters (β and capacity) are seen to change. Figure 6 illustrates how these two different types of distribution of ligand through the gel affect the capacity of the matrix.

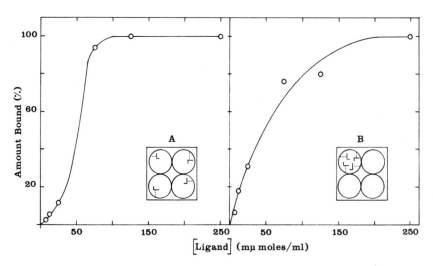

Fig. 6. Effect of ligand (L) distribution on the capacity of N^6-AMP-Sepharose for LDH-M_4.
    (A) Ligand uniformly coupled to Sepharose.
    (B) Ligand diluted with unsubstituted Sepharose.

The data in Table 2 for glyceraldehyde-3-phosphate dehydrogenase (G3PDH) support a qualitatively different interaction; only the phosphate immobilised adenylic acid binds this enzyme. This suggests a much more critical binding site for the adenine moiety. The work of Hocking and Harris (25) could be in agreement with this observation: C-8 immobilized NAD binds the enzyme well. This suggests that C-8 immobilized AMP should bind G3PDH also. As more detailed maps of dehydrogenase structure are published we should be able to investigate the feasibility of this type of probe of enzyme-ligand interaction. Additional implications can be drawn from the data in Table 2; the phosphate group is essential to the binding of yeast ADH and glycerokinase (GK). It has a quite different role in the interaction of the nucleotide and myokinase (or G3PDH) where the adenine moiety appears to be the vital portion of the nucleotide.

These results should be treated cautiously, since the binding of MDH to P^2-immobilised ADP was apparently tighter than that to N^6-AMP when the concentration of the latter was one-fourth that of the phosphate immobilized ligand. By increasing the concentration of N^6-AMP to 4μmoles/ml an increase in β was observed (Fig. 7). A similar explanation could explain the apparent differences for 3-phosphoglycerate and pyruvate kinases on these two adsorbents.

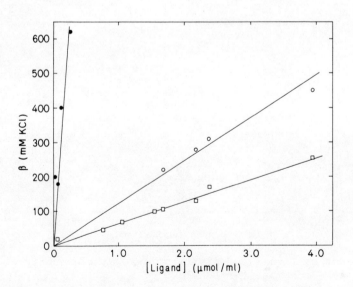

Fig. 7. Effect of N^6-AMP concentration on binding (β). (●) LDH-H_4; (○) MDH; (□) GK.

Similarly, the NADP-dependent glucose-6-phosphate DH binds preferentially to P²-ADP at pH 7.5. It can be retarded on N⁶-AMP by suitably adjusting the pH of the application buffer (Fig. 10).

The interaction of myokinase with P²-ADP has been studied over a range of concentrations by diluting with unsubstituted gel. The results are shown in Fig. 8.

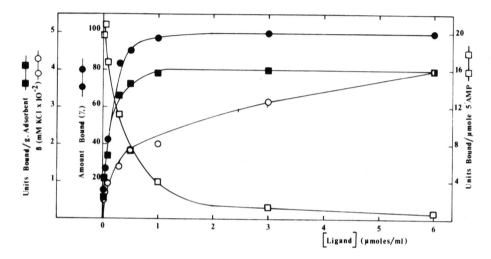

Fig. 8. The effect of P²-ADP concentration on the binding parameters of myokinase.
(•) % Bound   (o) β (mM KCl)
(□) Units bound per μmole ligand.
(■) Units bound per g. adsorbent.

β decreases as expected and agrees with parallel experiments with LDH and GK. In addition, however, the capacity of this adsorbent can be expressed as units bound per mole of ligand; interestingly, capacity is seen to increase as the ligand is diluted. This effect is qualitatively independent of both the nature of the adsorbent and the enzyme. Quantitative differences exist, however, for example, between the two LDH isoenzymes and GK. The interaction of the former is stronger for N⁶ alkyl-AMP than that of GK.

Finally, it should be noted that comparisons between P²-ADP and N⁶-AMP adenylic acids are difficult to make because of charge differences between the two polymers (the extra phosphate could complicate the interpretation of the phosphate-immobilised nucleotide.

Table 3 compares the relative behaviour of a variety of enzymes on three matrices: 1) aminoalkyl-NAD, 2) aminoalkyl-AMP, both defined polymers and 3) aminoalkanoyl-NAD, an undefined polymer.

Several interesting features deserve comment: LDH binds more tightly to alkyl-NAD than alkanoyl-NAD, as indeed do all of the NAD-linked enzymes (with the possible exception of β-hydroxybutyrate dehydrogenase, the value for which agrees with the above conclusion at lower ligand concentrations). This suggests that a more specific binding is obtained for alkylated nucleotides than for acylated ones. This may be due to further withdrawal of the N_6-amino group lone pair (it is already deactivated by the purine ring). The conclusion is that better group-specific affinity matrices for NAD-dependent dehydrogenases are obtained via alkyl linkages at N^6.

TABLE 3. A comparison of the binding of various enzymes to immobilised nucleotide affinity columns.

The immobilised ligand concentration was 2.0μmoles nucleotide/ml for each polymer.
I N^6-(6-aminohexyl)-NAD^+-Sepharose
II N^6-(6-aminohexyl)-5'-AMP-Sepharose
III ε-aminohexanoyl-NAD^+-Sepharose

Enzyme	Binding (β)-mM KCl		
	I	II	III
Lactate dehydrogenase (H_4)	0.35(a)	>1M(b)	325
Lactate dehydrogenase (M_4)	0.80(a)	>1M(b)	500
Alcohol dehydrogenase (yeast)	700	400(d)	415
Alcohol dehydrogenase (liver)	290	450(d)	130
D-glucose 6-P dehydrogenase	140	0	125
L-malate dehydrogenase	300	65	45
Isocitrate dehydrogenase	0	0(d)	220
D-3-hydroxybutyrate dehydrogenase	185(a)	250(d)	250(b)
D-glyceraldehyde 3-P dehydrogenase	1.0	0	>1M(b)
L-glutamate dehydrogenase	430	(c)	(c)
Glutathione reductase	75	RT	120
Hexokinase	0	0	100
Glycerokinase	RT	122	NT
Pyruvic kinase	100	100	0
3-phosphoglycerate kinase	NT	70	NT
Creatine kinase	NT	0	0
Myokinase	50	0	RT

(a) Elution could not be effected by 1M KCl, enzyme eluted in 0-5mM NADH gradient (10ml total volume).
(b) Elution was effected by a 200μl pulse of 5×10^{-3}M NADH.
(c) No enzyme recovered.
(d) β determined on a polymer containing 16εmoles 5'-AMP/ml.
RT Enzyme retarded by the polymer and eluted behind the void volume as determined by bovine serum albumin.
NT Not tested.

One must place some reservations on the interpretation of this data; because of the variation of β with ligand concentration and because of the unknown response of β with the latter (it is linear over the range 16-2,000 mμmoles/ml), also because of the undefined nature of the acylated NAD matrices, not all these β values are strictly comparable.

A very practical point arising out of Table 3 concerns the data for glutamate dehydrogenase (GDH). This enzyme has presented considerable problems in the past due to its habit of regularly disappearing on affinity columns. Up to this time we had been unable to displace the enzyme in an active form. However, using alkyl-NAD it can be seen that GDH binds quite tightly (430mM salt) and can be removed in reasonable yield.

We have used these defined polymers further; it is possible using group specific matrices to examine the physicochemical parameters of affinity chromatography and thereby to provide a framework on which to base recommendations for general procedures.

Fig. 9 shows the variation of binding of LDH to acyl-NAD and N^6-alkyl-AMP. Up to pH 8 there was no effect of pH on binding. Above this pH, apparent pKs of 8.5 and 9.7 were observed respective-

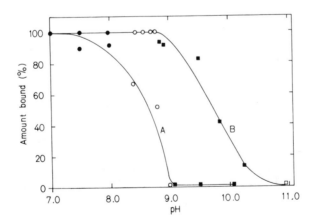

Fig. 9. The effect of pH on the binding of LDH-H_4 to (a) ε-aminohexanoyl-NAD^+-Sepharose and (b) N^6-(6-aminohexyl)5'-AMP Sepharose.
(●) 10mM KH_2PO_4-KOH   (o) 10mM Tricine-KOH   (■) 10mM Glycine-KOH

ly. The difference in the pK values might be a reflection of incomplete reaction of hexanoyl-agarose with NAD which would result in a residual charge on the polymer (other than nucleotide phosphate). In contrast, titration of N^6-alkyl-AMP only shows a plateau associated with the 5'-phosphate in the region of pH 6-7 (27). The pK measured for the interaction of LDH with the N^6-alkyl-AMP compares with the free solution pK which is also 9.7 (30). The lower pK found with the acyl-NAD polymer could be related to the modification of local hydrogen ion concentration by unsubstituted carboxyl groups.

Two proteins (GK and BSA) of identical overall charge to LDH were chromatographed under the same conditions. The BSA had no measurable affinity over the pH range studied whereas GK was shown to have an apparent pK of 8; and furthermore, GK was eluted at this pH with a pH gradient (6-10).

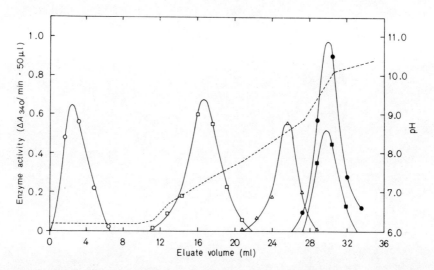

Fig. 10. The resolution of a dehydrogenase mixture on N^6-(6-aminohexyl) 5'-AMP-Sepharose by a pH gradient.
(O) BSA (●) MDH (□) G6PDH (■) LDH-H_4 (▲) YADH (----) pH

Complex mixtures of dehydrogenases may be separated on N^6-alkyl AMP using pH gradients (Fig. 10). G6PDH is adsorbed at 6.5 and eluted at 7.2. Advantages are seen by trying gradients running both from low to high pH's and vice versa. LDH like GK is eluted at the pK determined above. LDH is not normally eluted with salt but by nucleotide pulses. Hence pH gradients are useful when enzyme is required free of nucleotide, for example prior to kinetic work.

The pH dependence of the nucleotide enzyme interaction invariably does not follow changes in overall protein charge thereby broadening the scope of group specific affinity chromatography. It is likely to be particularly useful for very tightly bound enzymes.

THE EFFECTS OF TEMPERATURE ON GROUP SPECIFIC ADSORBENT CHROMATOGRAPHY

The marked changes of β with temperature were studied at two different ligand concentrations. At 1.5 μmoles/ml the capacity of N^6-alkyl AMP for YADH & GK was lowered with increasing temperature (Fig. 11). At the higher ligand concentration (4μmoles/ml) GK was quantitatively retained. At both ligand concentrations, the binding β was reduced at increased temperatures. Prolonged exposure of such columns to elevated temperatures does not alter their subsequent chromatographic properties at lower temperatures, and these temperature-dependent phenomena were not the result of a cumulative temperature-dependent deterioration of the adsorbent.

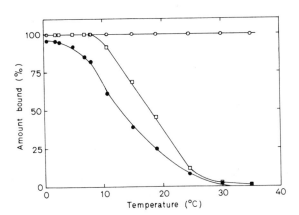

Fig. 11. The effect of temperature on the capacity of N^6-(6-aminohexyl) 5'-AMP-Sepharose.
(●) GK (□) YADH......Ligand concentration 1.5μmoles/ml.
(○) GK.....Ligand concentration 4.0μmoles/ml.

The binding of LDH to N^6-alkyl-AMP was sufficiently strong that even at 40°C, 1M KCl would not elute the enzyme. The temperature dependence of the interaction of this enzyme can be studied using NADH gradients (Fig. 12A).

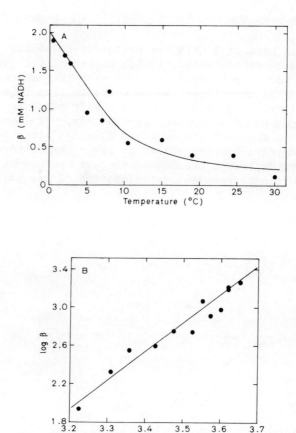

Fig. 12. The binding of LDH-H_4 to N^6-(6-aminohexyl) 5'-AMP-Sepharose in response to temperature.

The data from this experiment were replotted as an Arrhenius plot (Fig. 12) and was characteristic of an adsorption energy of 13 Kcal/mole. Free solution measurements of LDH and NADH indicate that the affinity decreases with increasing temperature. Therefore the effectiveness of the eluant NADH should decrease with temperature and this adsorption energy value is a minimum value.

By varying the temperatures of both the adsorption and desorption phases, it is possible to study this property of affinity systems further. The effect on the binding of GK by adsorbing the enzyme at 4°C and eluting at 23° and vice versa are shown in Fig. 13. Again it is seen that temperature decreases the enzyme's affinity for the ligand.

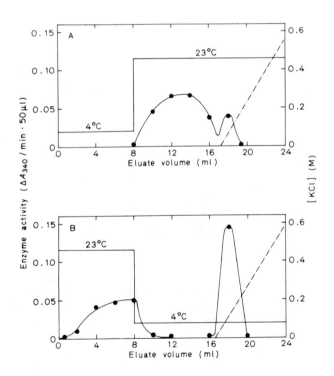

Fig. 13. The effect of temperature on the adsorption of GK to N^6-(6-aminohexyl)5'-AMP-Sepharose.

Figure 14 shows how four enzymes may be resolved on N^6-alkyl-AMP using a linear temperature gradient; these enzymes were YADH, GK, HK and LDH. The first two enzymes (GK and YADH) were eluted in the order expected from the adsorption energies determined as described above. LDH was recovered at 40° using a 5mM pulse of NADH.

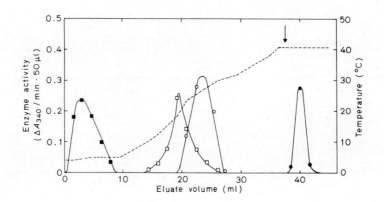

Fig. 14. The resolution of an enzyme mixture of N^6-(6-aminohexyl) 5'-AMP-Sepharose by a temperature gradient.
(■) HK (□) GK (○) YADH (●) LDH-H_4 (-----) °C ↓ NADH pulse

This experiment shows that when combined with other techniques (such as NADH pulses, salt gradients, lowering of the ligand concentration) temperature effects can be used to elute tightly bound enzymes and to greatly increase the range of separations of which a particular matrix is capable. An additional bonus is that the enzyme does not need to be freed from salt or nucleotide following desorption.

Temperature is one of the most important variables to be carefully controlled during chromatography and which has very dramatic effects on reproducibility of results: this effect is particularly marked when ionic gradients are used.

STUDIES OF COLUMN GEOMETRY AND DYNAMICS FOR N^6-ALKYL-AMP

The linear relationship between β and ligand concentration (Fig. 7) could be associated with the total amount of ligand rather than its concentration *per se*. When we kept both of these parameters

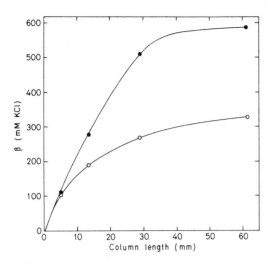

Fig. 15. The effect of column length, at constant ligand concentration and content, on the binding parameter β.

constant and varied the gel bed length by 60-fold and the diameter 4-fold, β approached a limiting value as the length increased (Fig. 15). However, the binding was inversely related to column length when the total amount of ligand used and column diameter were kept constant (the ligand concentration was varied inversely with the length, Fig. 16). A linear relationship was found between β and column length when the total amount of ligand at a fixed concentration was proportional to the column length (Fig. 17).

Thus we conclude that the strength of the interaction, whilst primarily dependent on ligand concentration is also dependent on the total amount of ligand and bed geometry. The effect of the last parameters is prevalent at low ligand concentrations.

Furthermore, of the above parameters only the total amount of ligand affected the capacity. There was no effect on the capacity of varying the enzyme concentration (up to 18 units per ml) under

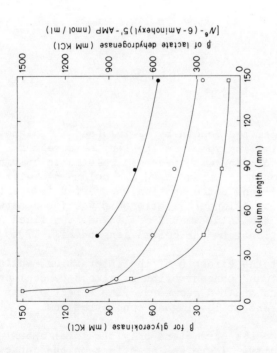

Fig. 16. The effect of column length, at constant ligand content, on the binding (β).

(○) GK (●) LDH-H$_4$
(□) N^6-AMP concentration

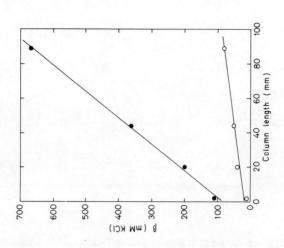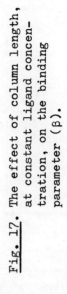

Fig. 17. The effect of column length, at constant ligand concentration, on the binding parameter (β).

(●) LDH-H$_4$ (○) GK

column conditions, however, under batchwise conditions the effect of enzyme concentration on capacity was a time-dependent process (pronounced at low enzyme concentrations) (Fig. 18). Similar effects can be induced under column conditions by reducing the ligand concentration, reflecting the time required to establish equilibrium. Under batchwise conditions, the affinities of GK and LDH for N^6-alkyl-AMP are comparable (26) even though their binding constants, measured under column conditions, differ by some 40-fold (Fig. 7). We conclude that the adsorption and desorption phases of affinity chromatography are two distinct processes; the release of the enzyme probably being mediated by an eluant-induced conformation change.

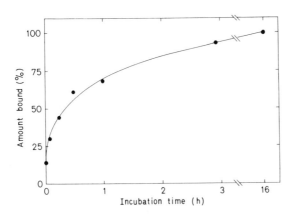

Fig. 18. The capacity of N^6-(6-aminohexyl) 5'-AMP-Sepharose for LDH-H_4 under batchwise conditions

THE EFFECTS OF FREE SOLUTION KINETICS ON THE BINDING OF ENZYMES TO N^6-(6-AMINOHEXYL) 5'-AMP-SEPHAROSE.

The dissociation constant (K_L) of LDH for N^6-alkyl-AMP was determined under batchwise conditions (Fig. 19, K_M = 1.1 x 10^{-6}M), whereas under column conditions the K_L approached 2.7 x 10^{-10}M (29). The difference between these two K_L values may be associated with the deviation under column conditions from the freely reversible equilibrium state. In both instances the increased affinity for N^6-alkyl-AMP upon immobilisation contradicts earlier postulates that ligand immobilisation impairs the interaction of the complimentary macromolecule (31,32). However, recent observations (33,34)

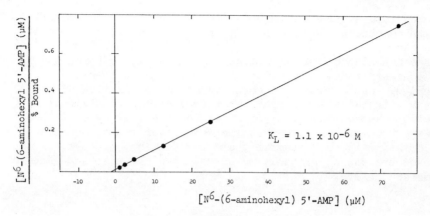

Fig. 19. The dissociation constant for LDH-H$_4$ on N^6-(6-aminohexyl) 5'-AMP-Sepharose under batchwise conditions.

suggest that this increased affinity may be the result of adding a hydrophobic extension arm, and thus the recorded affinities could be compounded from the intrinsic affinity for the ligand <u>plus</u> a hydrophobic term. In this context we have previously shown that LDH has a greater affinity for acyl-NAD when linked via hydrocarbon arms when compared with the same ligand linked via hydrophilic (polyglycine) arms (9). The affinities of several enzymes for N^6-alkyl-AMP change with the extension arm length as shown in Figure 20 (22).

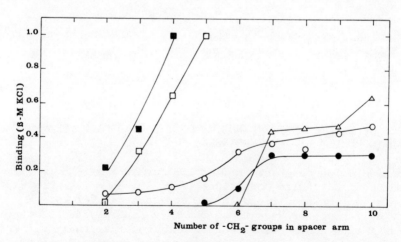

Fig. 20. The binding (β) of several enzymes to N^6-alkyl-AMP-Sepharose in response to extension arm length.
(□) LDH-H$_4$ (■) LDH-M$_4$ (●) G6PDH   (○) MDH   (△) G3PDH

Under column and batchwise conditions, LDH can be bound to N^6-alkyl-AMP in the presence of competing inhibitor at concentrations far exceeding those required to effect elution under normal chromatographic conditions, although as the concentration of inhibitor is increased the β value decreases (Fig. 21). This phenomenon supports the concept that the adsorption and desorption phases of this and probably other types of affinity chromatography are two distinct processes. Furthermore, enzymes are eluted according to their free solution affinities rather than their affinity for the immobilised ligand (6).

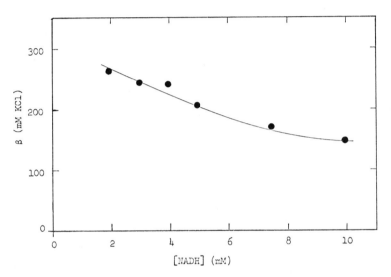

Fig. 21. The effect of NADH concentration on the binding of LDH-H_4 to N^6-(6-aminohexyl) 5'-AMP-Sepharose.

The behaviour of LDH in response to ligand concentration (24,26) and the partition of enzyme down a column equilibrated with free unimmobilised nucleotide (29) indicates that interaction with more than one ligand is required to retard the enzyme. Thus it is concluded that, in accord with Cuatrecasas, the enzyme is concentrated within the column by interacting with the immobilised ligand and this results in the tight adsorption of the enzyme. Under batchwise conditions the enzyme concentration would not be increased and therefore a lower ligand-enzyme affinity is recorded. These results have been used to propose that more than one ligand is required for binding (24,26,29) of one molecule of enzyme. Because under column conditions only 0.1% of the total ligand is used to bind enzyme, this percentage may indicate what fraction of the ligand is available in the conformation peculiar to the active site of the enzyme.

ACKNOWLEDGEMENTS

The authors would like to thank the Science Research Council (U.K.) for financial support; M.C. Hipwell for unpublished data; the European Journal of Biochemistry and FEBS Letters for permission to use many of the figures from references 20, 21, 24, 26-29.

REFERENCES

1. C. R. Lowe & P. D. G. Dean (1971) FEBS Lett., 14, 313.
2. R. Collier & G. Kohlhaw (1971) Anal. Biochem., 42, 48.
3. P. C. H. Newbold & N. G. L. Harding (1971) Biochem. J. 124, 1.
4. C. R. Lowe, K. Mosbach & P. D. G. Dean (1972) Biochem. Biophys. Res. Commun. 48, 1004.
5. K. Mosbach, H. Guilford, R. Ohlsson & M. Scott (1972) Biochem. J. 127, 625.
6. R. Ohlsson, P. Brodelius & K. Mosbach (1972) FEBS Lett., 25, 234.
7. C. R. Lowe & P. D. G. Dean (1973) Biochem. J., 133, 515.
8. C. R. Lowe, M. J. Harvey, D. B. Craven, M. A. Kerfoot, M. E. Hollows & P. D. G. Dean (1973) Biochem. J., 133, 507.
9. C. R. Lowe, M. J. Harvey, D. B. Craven & P. D. G. Dean (1973) Biochem. J. 133, 499.
10. R. H. Allen & P. W. Majerus (1972) J. Biol. Chem., 247, 7695.
11. C. R. Lowe & P. D. G. Dean (1971) FEBS Lett., 18, 31.
12. P. -O. Larsson & K. Mosbach (1971) Biotechnol. Bioeng. 13, 393.
13. P. D. G. Dean & C. R. Lowe (1972) Biochem. J. 127, 11P.
14. M. K. Weibel, H. H. Weetall & H. J. Bright (1971) Biochem. Biophys. Res. Commun. 44, 347.
15. J. D. Hocking & J. I. Harris (1972) Biochem. J. 130, 24P.
16. H. Guilford, P. -O. Larsson & K. Mosbach (1972) Chem. Scripta 2, 165.
17. R. Barker, K. W. Olsen, J. H. Shaper & R. L. Hill (1972) J. Biol. Chem. 247, 7135.
18. O. Berglund & F. Eckstein (1972) Eur. J. Biochem. 28, 492.
19. R. Lamed, Y. Levin & M. Wilchek (1973) Biochim. Biophys. Acta, 304, 231.
20. D. B. Craven, M. J. Harvey, C. R. Lowe & P. D. G. Dean (1974) Eur. J. Biochem. 41, 329.
21. D. B. Craven, M. J. Harvey & P. D. G. Dean (1974) FEBS Lett. 38, 320.
22. M. C. Hipwell & P. D. G. Dean (manuscript in preparation).
23. P. O'Carra & S. Barry (1972) FEBS Lett. 21, 281.
24. M. J. Harvey, C. R. Lowe, D. B. Craven & P. D. G. Dean (1974) Eur. J. Biochem. 41, 335.
25. J. D. Hocking & J. I. Harris (1973) FEBS Lett.

26. C. R. Lowe, M. J. Harvey & P. D. G. Dean (1974) Eur. J. Biochem. 41, 341.
27. C. R. Lowe, M. J. Harvey & P. D. G. Dean (1974) Eur. J. Biochem. 41, 347.
28. M. J. Harvey, C. R. Lowe & P. D. G. Dean (1974) Eur. J. Biochem. 41, 353.
29. C. R. Lowe, M. J. Harvey & P. D. G. Dean (1973) Eur. J. Biochem. (in press) VI.
30. A. D. Winer & G. W. Schwert (1958) J. Biol. Chem. 231, 1065.
31. P. Cuatrecasas & C. B. Anfinsen (1971) Annu. Rev. Biochem. 40, 259.
32. P. Cuatrecasas (1972). Advan. Enzymol. 36, 29.
33. B. H. J. Hofstee & N. F. Otillio (1973) Biochem. Biophys. Res. Commun. 53, 1137.
34. P. O'Carra, S. Barry & T. Griffin (1973). Biochem. Soc. Trans. 1, 289.

AFFINITY CHROMATOGRAPHY OF KINASES AND DEHYDROGENASES ON SEPHADEX®
AND SEPHAROSE® DYE DERIVATIVES

Richard L. Easterday and Inger M. Easterday

Pharmacia Fine Chemicals Inc.

800 Centennial Ave., Piscataway, N.J. 08854

Biospecific affinity chromatography has been shown to be a powerful method for the isolation of biological compounds[1,2]. While high specificity of the immobilized ligand for the compound of interest generally results in high purification factors and simple isolation protocols, it usually necessitates the often laborious synthesis of a special affinity adsorbent. Affinity gels with group rather than compound specificity therefore have the potential advantage of being useful for the isolation of many compounds belonging to a particular group. The resolving power of these gels need not be less than the more compound specific gels, since highly selective compound elution is still possible with substrates, cofactors, or inhibitors.

A number of workers have reported the binding of Blue Dextran 2000® to proteins such as phosphofructokinase[3,4], pyruvate kinase[5,6,7], glutathione reductase[8], acetoacetatesuccinyl CoA transferase[9], blood coagulation factors[10], and sweet corn R-enzyme[11]. Gel filtration of Blue Dextran 2000-protein complexes and then of dissociated complexes has been used as a purification procedure for a number of the above mentioned enzymes. Bohme et al[12] have also purified phosphofructokinase on Blue Dextran immobilized in polyacrylamide and Cibacron blue coupled to Sephadex G-200.

We have synthesized a number of new affinity gels by coupling the blue dye used in producing Blue Dextran 2000 to various Sephadex and cross-linked Sepharose gels. We have also studied the properties and usefulness of these gels for the isolation of certain kinases and dehydrogenases from rabbit muscle and yeast.

Blue Sephadex G-50 and G-200 were synthesized by coupling

Cibacron blue F3GA to Sephadex essentially according to the procedure described by Böhme et al[12]. Sepharose 2B and 6B were first cross-linked with epichlorohydrin and desulfated essentially according to the procedure of Porath, Janson, and Lääs[13] in order to increase their chemical stability before coupling of the dye. The synthesis and structure of Blue Sepharose 6B is shown in Figure 1. After coupling, all gels were washed repeatedly and extensively at pH 4.5 and 9 with 0.02 M acetate buffer in 2 M NaCl and 0.02 M bicarbonate buffer in 2 M NaCl respectively. After washing at pH 9, gels were washed with deionized water.

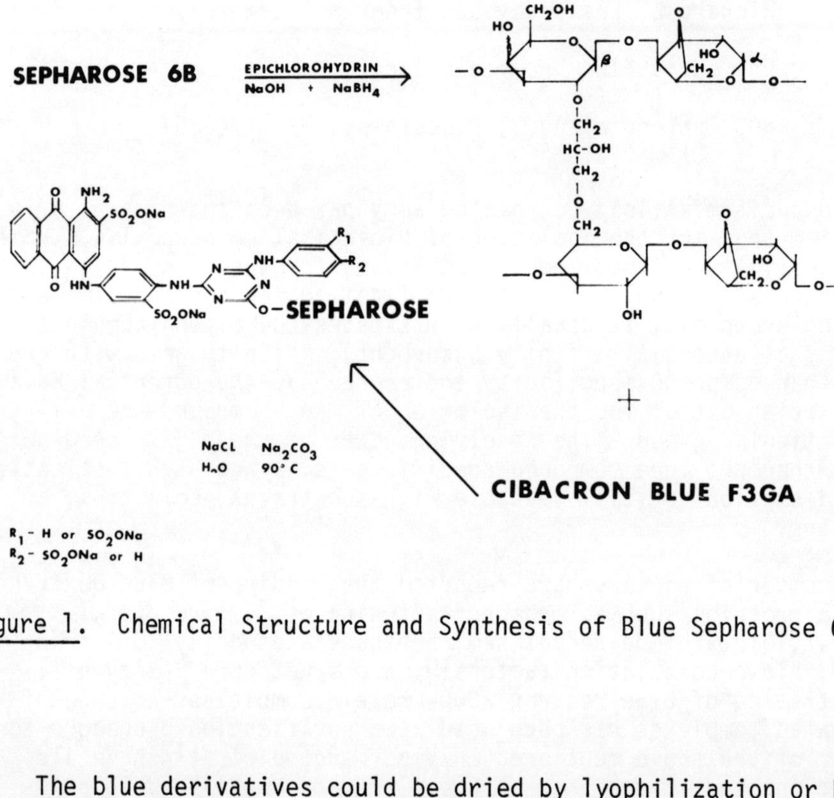

Figure 1. Chemical Structure and Synthesis of Blue Sepharose 6B

The blue derivatives could be dried by lyophilization or by washing with increasing concentrations of ethanol (50-100%) aspiration, and heating in a vacuum oven for 6 hrs. at 60°C. The dried adsorbents were easily rehydrated and showed no change in properties from undried gels.

The dye substitution was determined by drying the gels, hydrolyzing in 6 N HCl at 40°C for 30 minutes, reading the absorbance at 650 nm, and comparing the values to a standard curve prepared with pure dye. The results are shown in Table 1. The higher values for the Sephadex derivatives on a dry weight basis might be expected

Table 1.

Dye Substitution of Blue Sephadex and Sepharose Gels

Derivative	Dye Substitution	
	µmoles/g	µmoles/ml
Blue Sepharose 2B	79.5	0.75
Blue Sepharose 6B	71.4	2.1
Blue Sephadex G-200	180.0	4.0
Blue Sephadex G-50	118.0	8.0
Blue Dextran 2,000	99.3	-

since dextran has a larger number of free hydroxyl groups available for reaction than does agarose. The dye substitution for Blue Dextran 2000 is given for comparison.

Comparison studies of the binding of creatine kinase (CK) and pyruvate kinase (PK) to Blue Sephadex and Sepharose derivatives are shown in Figures 2 and 3. CK was eluted with a pH shift from 5.9 to 8.0 and PK was eluted with a linear 0.01-0.6 M KCl gradient. It appears that the Blue Sephadex G-50 is not sufficiently porous to allow good binding of the high molecular weight PK, but does bind the relatively low molecular weight CK quite well. The Blue Sephadex G-200 appeared to bind both enzymes quite well but suffered from the disadvantages of broad elution peaks and significant changes in bed volume with changing ionic strength. Blue Sepharose 6B gave good binding of both PK and CK and relatively narrow elution zones. The lower degree of dye substitution on the 6B derivative compared to the G-200 derivative probably accounts for the ease of elution of PK and CK from the Blue Sepharose 6B. At low ionic strength, the Blue G-200 also had a greater swelling factor than Blue 6B, 45 ml/g and 30 ml/g respectively. Blue Sepharose 6B also did not change bed volume with varying ionic strength, as did the Blue Sephadex G-200. The highly porous Blue 2B showed even tighter binding of PK and CK than the Blue 6B or Blue G-200. It was concluded that the Blue 6B had the best all around properties and further studies were conducted with this gel.

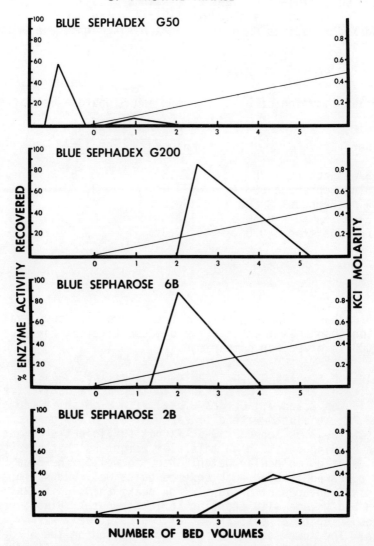

Figure 2. Purified pyruvate kinase (5 mg) was buffer exchanged on Sephadex G-25 and applied to 1.6 x 6 cm beds of Blue Sepharose 2B, Blue Sepharose 6B, Blue Sephadex G-200, and Blue Sephadex G-50. Starting buffer was 0.1 M citrate/phosphate, pH 5.5; containing 5×10^{-3} M $MgCl_2$, 2×10^{-6} M 2-mercaptoethanol, 4×10^{-4} M EDTA and 0.01 M KCl. Columns were eluted at a flow rate of 30 cm/hr with a 0.01-0.6 M linear KCl gradient having a volume equal to six bed volumes. Note that only fractions with peak activity and those where activity began and ended have been plotted.

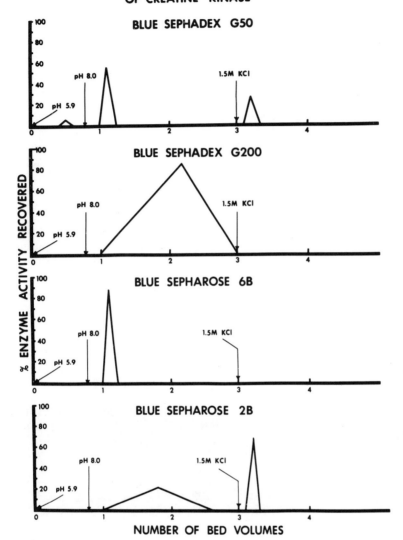

Figure 3. Purified creatine kinase (5 mg) was buffer exchanged on Sephadex G-25 and applied to 1.6 x 6 cm beds of Blue Sephadex G-50, Blue Sephadex G-200, Blue Sepharose 6B, and Blue Sepharose 2B. Starting buffer was 0.01 M phosphate, pH 5.9; containing 2×10^{-6} M 2-mercaptoethanol, 4×10^{-4} M EDTA, 5×10^{-3} M $MgCl_2$. Columns were eluted as noted with starting buffer at pH 5.9, pH 8.0 and with 1.5 M KCl added. Note that only fractions with peak activity and those where activity began and ended have been plotted.

Since the blue chromophore contains both amino and sulfonic acid groups, nonbiospecific ion exchange effects would be expected. Even though biospecific elution can diminish the importance of nonspecific binding, the binding of a number of proteins to Blue Sepharose was investigated. A sample containing 15 mg of each protein was chromatographed on a 1.6 x 5 cm bed at a flow rate of 30 cm/hr. Each protein was dissolved in and applied in the following buffers, 0.02 M phosphate, pH 8.5; 0.02 M phosphate, pH 6.5. The proteins were eluted in starting buffer with a 0-1.0 M linear KCl gradient equal to 20 bed volumes. The amount of protein eluted in the breakthrough volume, the molarity of KCl required for the elution of material not in the breakthrough volume, and the protein recovered by KCl elution are shown in Table 2. Although ovalbumin, bacitracin, thyroglobulin, and hemoglobin (except at low pH) did not show significant binding at low ionic strength, ionic and pH dependent binding of proteins such as chymotrypsinogen A, aldolase, and serum albumin did occur. The binding capacity for serum albumin, a protein known for its ability to bind dyes, was found to be 28.9, 40.8, and 32 mg/ml at pH 8.4 in phosphate, pH 6.5 in phosphate, and pH 8.4 in Tris HCl buffers respectively. Travis and Pannell[14] have suggested the use of a similar derivative (Blue Dextran 2000 coupled to Sepharose 4B) for the removal of albumin from plasma or serum.

Table 2.

Nonspecific Protein Binding to Blue Sepharose 6B

Protein	0.02 M PO$_4$ Buffer pH	Protein Unbound %	KCl for Elution (M)	KCl Eluted Protein %
Ovalbumin	6.5	97.0	-	-
"	8.5	99.5	-	-
Bacitracin	6.5	92.0	-	-
"	8.5	92.4	-	-
Cytochrome C	6.5	7.9	0.27	91.2
"	8.5	12.1	0.23	87.9
Chymotrypsinogen A	6.5	3.1	0.22	96.9
"	8.5	7.1	0.22	92.9
Aldolase	6.5	-	0.38	95.2
"	8.5	-	0.34	98.3
Thyroglobulin	6.5	90.5	-	-
"	8.5	92.7	-	-
Hemoglobulin	6.5	4.8	0.28	95.2
"	8.5	82.2	1.0	14.3
Albumin	6.5	-	0.54	97.0
"	8.5	-	1.0	99.7

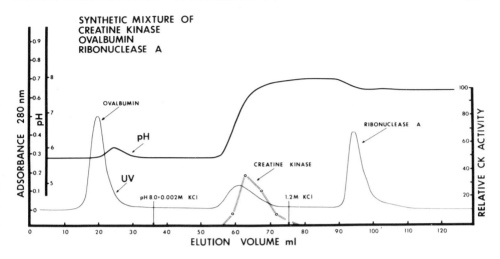

Figure 4. A mixture containing 5 mg each of creatine kinase, ovalbumin, and ribonuclease A was dissolved in starting buffer and applied to a 1.6 x 6 cm bed of Blue Sepharose 6B. The column was equilibrated at 30 cm/hr and eluted as noted with starting buffer (S.B.) consisting of 0.01 M phosphate, pH 5.8; 2×10^{-6} M 2-mercaptoethanol; 4×10^{-4} M EDTA; 5×10^{-3} M $MgCl_2$. Creatine kinase was eluted with starting buffer at pH 8.0 containing 1×10^{-3} M KCl, and ribonuclease A was eluted with starting buffer pH 8.0 containing 1.2 M KCl. Absorbance and pH were monitored with a Pharmacia Duo Monitor at 280 nm and Markson flow-through electrode respectively.

The separation of a synthetic mixture of creatine kinase, ovalbumin and ribonuclease A using pH and ionic strength changes is shown in Figure 4. Such changes in elution conditions can be used for the separation of simple mixtures and perhaps for more complex mixtures where the protein binding properties are appropriate. It can be convenient and advantageous since expensive cofactors or substrates are not required for elution. It is unlikely however that if the ligand does not have a high degree of specificity for the compound of interest, then a biospecific solute will be required for high resolution.

Biospecific elution of lactate dehydrogenase (LDH) and PK and elution of CK by pH change is shown in Figure 5. As can be seen from Table 3, good purification and recoveries with little cross-contamination of enzymes was achieved. It was also observed that LDH was not eluted with 10 mM NAD, as it was with 3 mM NADH, under otherwise identical conditions.

Table 3.

Purification Data for Creatine Kinase, Lactate Dehydrogenase, and Pyruvate Kinase on Blue Sepharose 6B

Peak	pH	Selective Eluant	Spec. Act. Increase	% Enzyme Act. Recovered		
				CK	LDH	PK
Breakthrough	6.9	Starting Buffer	-	<1	<1	0
CK	8.5	0.005 M KCl	5.2x	74.0	0	0
LDH	8.5	3 mM NADH	21.5x	0.4	72	0
PK	8.5	10 mM ADP	22.8x	0.8	0	64

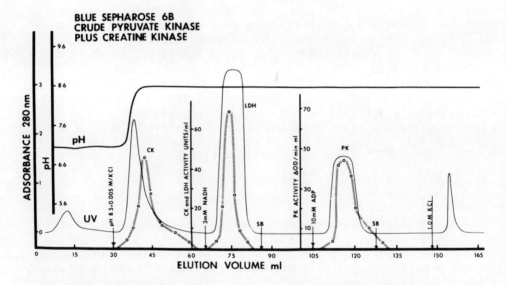

Figure 5. Partially purified rabbit muscle preparation containing approximately 15 mg of pyruvate kinase, 7.5 mg lactate dehydrogenase and 10 mg of added creatine kinase was buffer exchanged on Sephadex G-25 and then applied to a 1.6 x 5 cm bed of Blue Sepharose 6B. The column was equilibrated and eluted at 30 cm/hr as noted with starting buffer (S.B.) consisting of 0.02 M Tris HCl, pH 6.9; 5 x 10^{-3} $MgCl_2$; 4 x 10^{-4} M EDTA; 2 x 10^{-6} M 2-mercaptoethanol. Creatine kinase was eluted with starting buffer at pH 8.5 containing 5 x 10^{-3} M KCl and lactate dehydrogenase and pyruvate kinase were eluted with the same buffer plus 3 mM NADH and 10 mM ADP respectively. Additional UV adsorbing material was eluted with 1.0 M KCl.

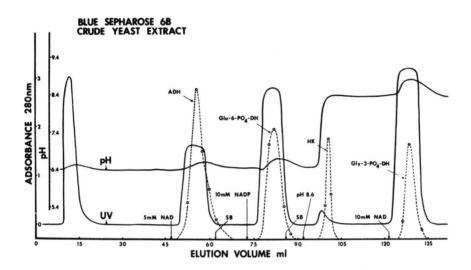

Figure 6. Dried bakers' yeast was lysed in 1 M dibasic sodium phosphate for 3 hours at 37°C and then centrifuged at 13,700 x g for 1 hour. The supernatant was filtered through cheesecloth and 0-75% saturation $(NH_4)_2SO_4$ cut was made at 4°C. The precipitate was recovered by centrifugation and then dissolved in starting buffer (S.B.) and buffer exchanged on Sephadex G-25. A 1.6 x 5 cm bed of Blue Sepharose 6B was equilibrated with starting buffer consisting of 0.02 M Tris HCl, pH 6.4; 5×10^{-3} $MgCl_2$; 4×10^{-4} M EDTA; 2×10^{-6} M 2-mercaptoethanol. A sample of crude extract containing 223 O.D. units in 10 ml was applied to the bed and washed with starting buffer. Various enzymes were eluted in starting buffer at pH 6.4 or 8.6, as noted, with the addition of selective cofactors. Alcohol dehydrogenase (ADH), glucose 6-phosphate dehydrogenase (glu-6-PO_4-DH), hexokinase (HK), and glyceraldehyde 3-phosphate dehydrogenase (gly-3-PO_4-DH) were eluted with 5 mM NAD, 10 mM NADP, pH 8.6, and 10 mM NAD respectively.

The purification of four enzymes from a crude yeast extract with high elution specificity is shown in Figure 6. Excellent purification factors of 31 to 4,460x, as shown in Table 4, were obtained with good recoveries, except in the case of hexokinase (HK). Elution conditions for HK were not optimal since 10% of the activity eluted in the breakthrough and 11% with the alcohol dehydrogenase activity. The use of ATP to elute HK caused serious leakage of other dehydrogenases. Yeast glucose 6-phosphate dehydrogenase is highly specific for NADP. The inability of NAD to elute glu-6-PO4DH under identical conditions where NADP eluted the enzyme with 60% recovery and 4,460-fold purification, illustrates the power of selective elution.

Table 4.

Purification Data for Separation of Yeast Enzymes on Blue Sepharose 6B

Peak	pH	Selective Eluant	Spec. Act. Incr.	% Enzyme Recovery			
				ADH	Glu-6-PO4DH	Hexokinase	Gly-3-PO4DH
Void				<1	0	10	0
ADH	6.4	10mM NAD	31x	55	0	11	0
Glu-6-PO4DH	6.4	10mM NADP	4,460x	0	60	0	0
Hexokinase	8.5	S. Buffer	10.2x	0	0	25	0
Gly-3-PO4DH	8.5	10mM NAD	155.0x	0	0	0	55

Blue Sepharose gels appear promising as general affinity matrices for both kinases and dehydrogenases. Selective elution with cofactors has been demonstrated, although the use of cofactor gradients needs investigation.

The chemical stability of these gels appears to be quite good and the ability to store them dry is very advantageous.

Additional studies are under way to compare the affinity properties of Blue Sepharose gels to AMP- and NAD-Sepharose.

References

1. P. Cuatrecasas and C.B. Anfinsen, Methods in Enzymology, 22, 345 (1971).
2. J. Porath and T. Kiristiansen, The Proteins, Vol. I, 3rd Ed. (1973).
3. G. Kopperschläger, R. Freyer, W. Diezel, and E. Hofmann, FEBS Letters, 1, 137 (1968).
4. G. Kopperschläger, W. Diezel, R. Freyer, S. Liebe, E. Hofmann, Eur. J. Biochem., 22, 40 (1971).
5. H. Haeckel, B. Hess, W. Lauterborn, and K.H. Wüster, Hoppe-Seyler's Z. Physiol. Chem., 349, 699 (1968).
6. K.G. Blume, R.W. Hoffbauer, D. Busch, H. Arnold, G.W. Löhr, Biochim. Biophys. Acta, 227, 364 (1971).
7. G. E. Staal, J.F. Koster, H. Kamp, L. Van-Milligen-Boersma, C. Veeger, Biochim. Biophys. Acta, 227, 86 (1971).
8. G.E. Staal, J. Visser, C. Veeger, Biochim. Biophys. Acta, 185, 39 (1969).
9. H.D. White and W.P. Jencks, Amer. Chem. Soc. Meeting Abstracts, Abstr. No. 43, (1970).
10. A.C.W. Swart and H.C. Hemker, Biochim, Biophys. Acta, 222, 692 (1970).
11. J.J. Marshall, J. Chromatogr., 53, 379 (1970).
12. H.J. Böhme, G. Kopperschläger, J. Schulz, and E. Hofmann, J. Chromatogr., 69, 209 (1972).
13. J. Porath, J.C. Janson, and T. Låås, J. Chromatogr., 60, 167 (1971).
14. J. Travis and R. Pannell, Clinica Chimica Acta (to be published)

AFFINITY CHROMATOGRAPHY OF THYMIDYLATE SYNTHETASES USING 5-FLUORO-2'-DEOXYURIDINE 5'-PHOSPHATE DERIVATIVES OF SEPHAROSE

John M. Whiteley, Ivanka Jerkunica and Thomas Deits

Scripps Clinic and Research Foundation

La Jolla, California 92037

The *de novo* synthesis of thymidylate from deoxyuridylate proceeds by the following sequence of reactions:

$$5,10\text{-Methylenetetrahydrofolate} + \text{deoxyuridylate} \longrightarrow \text{dihydrofolate} + \text{thymidylate} \quad (1)$$

$$\text{Dihydrofolate} + \text{TPNH} + \text{H}^+ \longrightarrow \text{tetrahydrofolate} + \text{TPN}^+ \quad (2)$$

$$\text{Tetrahydrofolate} + \text{"-CH}_2\text{-"} \longrightarrow 5,10\text{-methylenetetrahydrofolate} \quad (3)$$

In the presence of thymidylate synthetase, 5,10-methylenetetrahydrofolate donates its labile one-carbon unit (the methylene group) to deoxyuridylate, and also provides the reducing power necessary to convert the methylene group to methyl leading to the formation of thymidylate and dihydrofolate, as is shown in equation (1). The dihydrofolate so formed is converted to tetrahydrofolate with the aid of a second enzyme, dihydrofolate reductase (equation (2)). The cycle is then completed (equation (3)) by tetrahydrofolate acquiring the one-carbon unit necessary to re-form the methylene derivative utilized in the first reaction. Because thymidylate is essential for the biosynthesis of DNA, these reactions play a leading role in cellular replication. This factor is probably responsible for the potent chemotherapeutic properties possessed by compounds such as amethopterin and 5-fluorouracil. The former inhibits reaction (2) and a derivative of the latter inhibits reaction (1). The mechanisms of these three reactions have been the object of research in many laboratories during the past decade (1-4). The recent development of antifolate-resistant cell lines, which contain elevated levels of dihydrofolate reductase and thymidylate synthetase (5,6), has meant that these enzymes can be isolated in quantities sufficient to carry out precise structural studies such as sequencing, active-site labeling and X-ray crystallography.

The current structural and mechanistic studies on dihydrofolate reductase have been greatly aided by using affinity chromatography for the isolation of this enzyme (7,8). This observation suggested that a similar procedure might be equally appropriate for extracting thymidylate synthetase from crude cellular extracts. Although homogeneous samples of this enzyme had been isolated previously by conventional procedures, the operations were numerous and the overall yields were low (6,9). The use of antifolate-resistant bacterial cell lines, containing elevated enzyme levels, ensured product recovery despite sub-optimal conditions for isolation. However, from mammalian cells even sources such as regenerative liver tissue provide much lower levels of the synthetase[§], hence there is a strong requirement for a procedure which could ensure an efficient recovery of active protein.

It had been shown that fluorodeoxyuridylate was a good inhibitor of thymidylate synthetase (10). Hence, this molecule was incorporated by various procedures into a Sepharose matrix, and an amethopterin-resistant strain of *Lactobacillus casei*, a good source of the synthetase, was used to determine which column material was most suited for purification of the enzyme by affinity chromatography. The column materials which were prepared are shown in Fig. 1. The linking groups used to couple the inhibitor to Sepharose were: (1) aminohexyl; (2) 6-*p*-aminobenzamidohexyl; (3) 5'-*p*-aminophenyl; (4) 3'-*p*-aminophenylphosphate; and (5) aminohexyl/succinyl/*p*-aminophenyl.

Column material (1) was prepared by coupling diaminohexane to cyanogen bromide-activated Sepharose (11) and condensing the resultant aminohexyl derivative, in a carbodiimide-promoted reaction, with fluorodeoxyuridylate. To prepare derivative (2), *p*-nitrobenzoyl chloride was reacted with 6-aminohexane-1-ol in a procedure similar to that described by Danenberg *et al.* (12), and the product was coupled to fluorodeoxyuridylate in the presence of carbodiimide. The nitro group was catalytically reduced to an amino group, and the product was coupled in the usual way to activated Sepharose. For column materials (3) and (4), it was necessary to prepare the 5'- and 3'-*p*-aminophenyl esters of

---

Abbreviations: thymidylate, 2'-deoxythymidine 5'-phosphate; tetrahydrofolate, 5,6,7,8-tetrahydropteroylglutamic acid; dihydrofolate, 7,8-dihydropteroylglutamic acid; deoxyuridylate, 2'-deoxyuridine 5'-phosphate; fluorodeoxyuridine, 5-fluoro-2'-deoxyuridine; fluorodeoxyuridylate, 5-fluoro-2'-deoxyuridine 5'-phosphate; PMR, proton magnetic resonance.

[§] J. Galivan, personal communication.

AFFINITY CHROMATOGRAPHY OF THYMIDYLATE SYNTHETASES

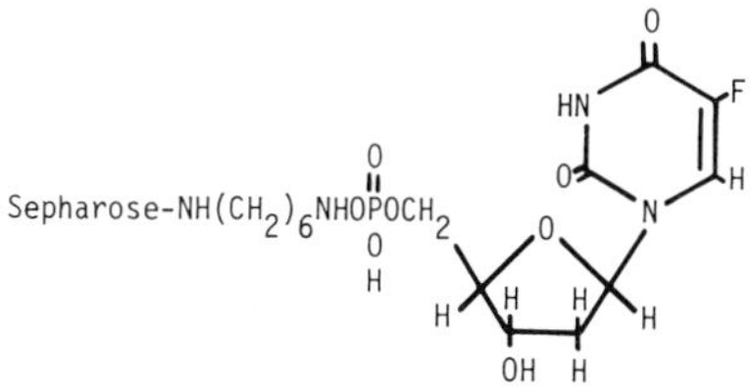

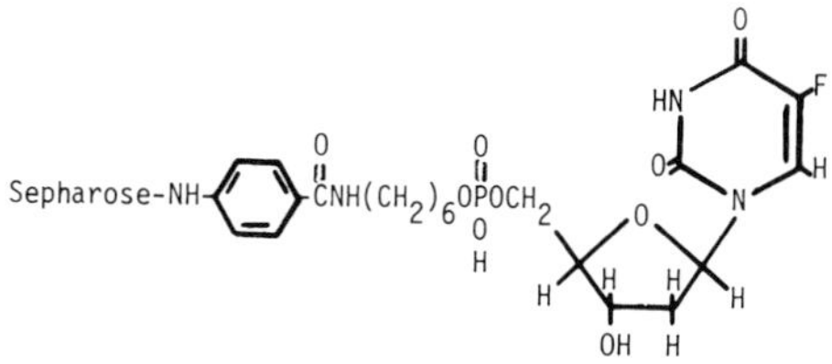

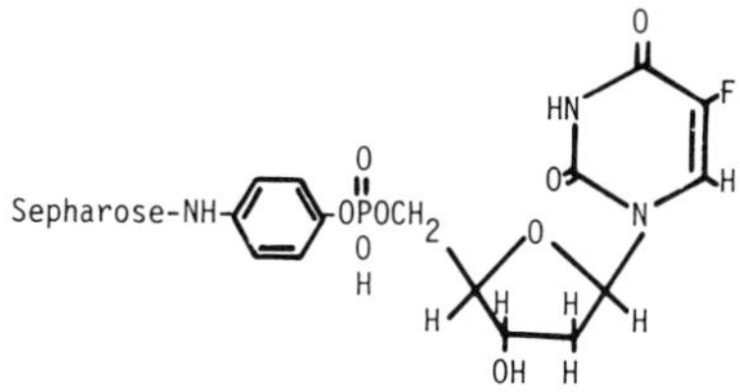

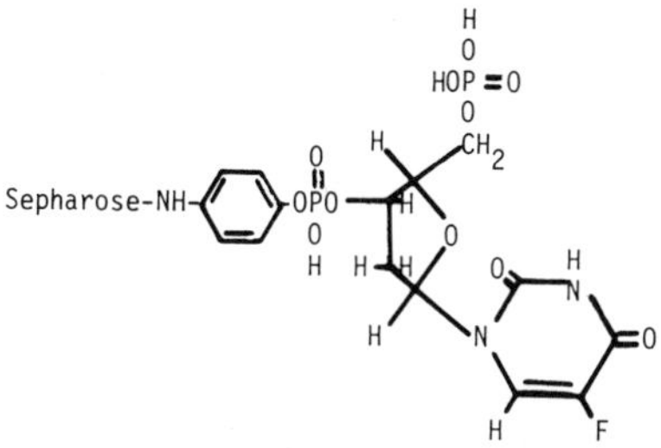

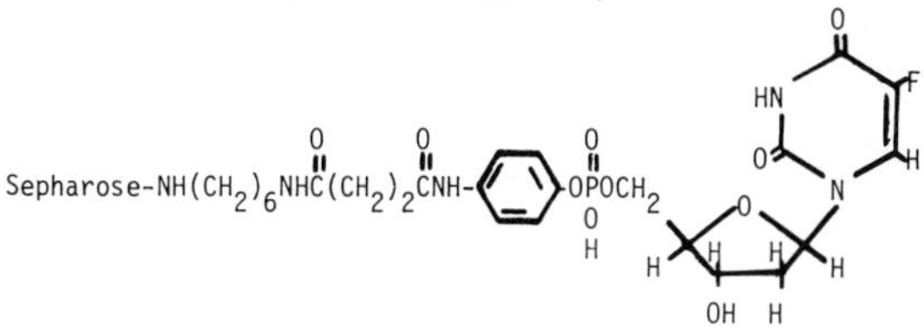

FIG. 1. Sepharose derivatives synthesized for use as affinity chromatographic column materials.

fluorodeoxyuridylate and fluorodeoxyuridine 3',5'-diphosphate (13), as outlined in Fig. 2. Fluorodeoxyuridine (a gift kindly supplied by Hoffman-La Roche) was converted in high yield by a carbodiimide-promoted reaction with 2-cyanoethylphosphate, followed by hydrolysis, to the 3',5'-diphosphate (14). The diesterification was achieved

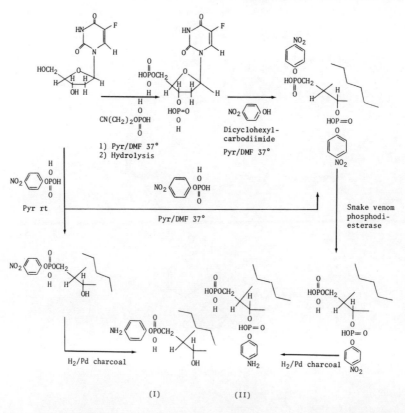

FIG. 2. The synthetic routes chosen to prepare the thymidylate synthetase inhibitors, 5-fluoro-2'-deoxyuridine 5'-p-aminophenylphosphate (I) and 5-fluoro-2'-deoxyuridine 3'-p-aminophenylphosphate 5'-phosphate (II).

in the presence of dimethylformamide at 37°. The level of substitution was monitored by PMR spectral measurements (Fig. 3) of the cyanoethylphosphates, which were isolated as their calcium salts prior to hydrolysis. The diphosphate was converted to the di-p-nitrophenyl ester by reaction with p-nitrophenol using reaction conditions similar to those in the first stage. The diester was also prepared by direct interaction of p-nitrophenylphosphate and 5-fluorodeoxyuridine. The 5'-ester linkage was cleaved with snake venom phosphodiesterase and the 3'-p-aminophenyl ester (Fig. 2, structure II) was obtained by reducing the remaining nitro- group with hydrogen and palladium charcoal (15). The 5'-mono ester (Fig. 2, structure I) was best obtained by reacting p-nitrophenylphosphate and 5-fluorodeoxyuridine under the less forcing conditions of pyridine alone as the solvent at room temperature. Both products I and II were isolated as their barium salts by fractional crystallization from ethanol:water mixtures. The free acids tended to

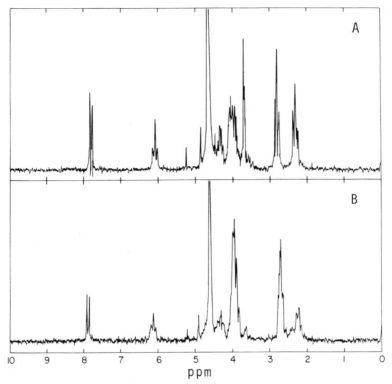

FIG. 3. The 100 MHz PMR spectra of the 2-cyanoethylphosphate derivatives of 5-fluoro-2'-deoxyuridine. Panel A, monoester; Panel B, diester. The spectra were taken in D_2O relative to hexamethyldisiloxane as internal standard. The multiplets at 2.8 and 4 ppm contain the protons assigned to the substituent ester groups.

rapidly decompose when concentrated or during filtration. The p-aminophenyl esters were then linked directly to activated Sepharose. In column material (3), the linkage is through the 5'- position, and in (4) it is through the 3'- position. For column material (5), the 5'-aminophenyl ester of fluorodeoxyuridylate was coupled in a carbodiimide-promoted reaction to succinylaminohexyl-Sepharose. An important variant of this column material was obtained when the excess succinyl carboxyl groups were blocked with glycinamide (16).

Table I presents a summary of the kinetic constants derived when the 5'- and 3'-p-aminophenyl phosphate derivatives (structures I and II) were allowed to interact with homogeneous samples of thymidylate synthetase. One enzyme was isolated from the amethopterin-resistant strain of *L. casei* developed by Dunlap and Harding (6), and the other from the T_2 phage-infected *Escherichia coli* system of Galivan and Maley. The former enzyme exhibited

TABLE I

Enzyme Source	Inhibitor	pH	K_i	K_m
Amethopterin-resistant *L. casei*	pdFU	7.2	8×10^{-9}	
	Aminophenyl-pdFU (I)	6.5	3.5×10^{-6}	
		7.2	8×10^{-6}	1.7×10^{-5}
		8.5	8.5×10^{-6}	
	pdFUp-aminophenyl (II)	7.2	6×10^{-4}	
T_2 Phage-infected *E. coli*	pdFU	7.2	1×10^{-8}	
	Aminophenyl-pdFU (I)	6.5	3×10^{-6}	
		7.2	5×10^{-6}	6×10^{-6}
		8.5	4×10^{-6}	
	pdFUp-aminophenyl (II)	7.2	6×10^{-5}	
		8.5	1×10^{-4}	

The inhibition constants shown by 5-fluoro-2'-deoxyuridine-5'-phosphate (pdFU), its 5'-*p*-aminophenyl ester (I) and its 3'-*p*-aminophenylphosphate derivative (II) for thymidylate synthetases from amethopterin-resistant *L. casei* and T_2 phage-infected *E. coli*. In each case, enzyme and inhibitor were pre-incubated for 10 min at 30° in the presence of 5,10-methylenetetrahydrofolate prior to the addition of substrate. The K_m values of the substrate, deoxyuridylate, are included for purposes of comparison.

maximal activity at pH 6.5-6.8, whereas the phage enzyme had a pH optimum of 8. The K_m values for the substrate, dUMP, determined at the intermediate pH value of 7.2, were 17 and 6×10^{-6} M, respectively. The K_i values were derived from the usual reciprocal plots and the esters, I and II, showed typical competitive inhibitory characteristics. When measured under the same conditions, the K_i for fluorodeoxyuridylate was $\sim 10^{-8}$ M for both enzymes. It can be seen in Table I that the 5'- derivative (I) inhibited both enzymes to a similar extent and that this inhibition varied little over the experimental range of pH. In contrast, the 3'- derivative (II) was a much poorer inhibitor. Representative inhibitor plots for compounds (I) and (II) at pH 7.2 are shown in Fig. 4. The substrate and inhibitor concentrations consistent with measurable reaction rates are indicated.

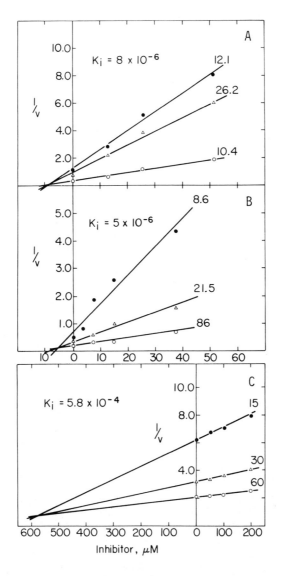

FIG. 4. Representative reciprocal plots used to derive the inhibition constants tabulated in Table I. (A) shows the inhibition of the *L. casei* synthetase by compound (I); (B) shows inhibition by the same compound for the enzyme from T_2 phage-infected *E. coli*; and (C) shows inhibition of the *L. casei* enzyme by compound (II). The μM concentration of the substrate, dUMP, is indicated for each curve. Reaction velocity, v, is expressed in nmoles/min.

Columns (ca. 1.5 x 10 cm) of the inhibitor-linked Sepharose materials were prepared, and their usefulness as affinity resins was evaluated by determining their ability to extract thymidylate synthetase from a cell-free extract of amethopterin-resistant *L. casei* (20 ml containing ca. 0.5 units/ml of enzyme[§]). Column materials (3) and (4) showed little ability to retain activity. In both cases, it was considered that this poor retention resulted primarily from the protein active site not seeing the inhibitor on the matrix because of the comparatively large size of the synthetase molecule (MW ~60,000) from this source (17), and that a more extended linking agent might overcome this problem. In the case of column material (4), however, a high inhibition constant determined for the 3'-p-aminophenylphosphate derivative (II) provided a second reason why significant interaction with the synthetase might be low for matrices containing the inhibitor bound directly through this position. Therefore, no further studies were made of the 3'-linked affinity matrices. In the case of column (1), enzyme activity was retained, but a comparison of inhibitor content (0.15 μM/ml), with the high level of aminohexyl content (86 μM/ml) suggested that retention was merely an ion-exchange process; this observation was reinforced by the recovery of only partially-purified enzyme.

Column material (2) was superior to each of the previous derivatives. In this case, the mode of preparation ensured that no excess potential ion-exchange groupings would be attached to the matrix. Unfortunately, the 5-fluoro-2'-deoxyuridine-5'-(6-p-aminobenzamidohexyl) phosphate was difficult to isolate as a homogeneous solid. In addition, a mixed solvent system of dimethylformamide/water was required in the cyanogen bromide-activated coupling of inhibitor to the polysaccharide matrix. When a column of this material was eluted with dimethylformamide, a continuous elution of inhibitor occurred, suggesting that non-specific adhesion, rather than true coupling, had taken place. However, at an arbitrary point elution with organic solvent was terminated and a column material was prepared for enzyme purification by elution with 0.01 M phosphate buffer, pH 7[†]. At this point, the Sepharose contained 6 μmoles/ml of inhibitor. Cell-free extract (10 ml, containing synthetase of specific activity 0.064) was applied to the column. The buffer strength was increased in a stepwise manner and, with 0.2 M phosphate, a quantitative recovery of enzymatic activity was achieved; unfortunately, the specific activity was only 0.66, corresponding to a mere ten-fold purifica-

---

[§] One unit of enzyme activity is defined as the amount required to synthesize 1 μmole of thymidylate/min under the experimental conditions. Specific activity is expressed as units/mg protein.

[†] All buffer solutions contained 0.02 M β-mercaptoethanol.

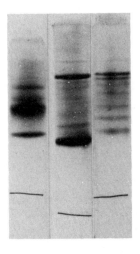

A B C

FIG. 5. Electrophoresis of samples of thymidylate synthetase on polyacrylamide gels. Aliquots were taken from: (A) leading edge; (B) peak; and (C) trailing edge of the principal enzymatically-active protein peak eluted by 0.2 M phosphate buffer, pH 7, from column material (5)(cf. Fig. 1). Electrophoresis was performed on 7.5% w/v acrylamide containing 0.2% w/v methylenebisacrylamide. Protein in each case migrated toward the positive electrode. The origin can be seen as a break in gel continuity at the top of each tube, and the front is marked by a nylon bristle insert at the base.

tion. It was considered that the difficulties previously described in purifying the inhibitor derivative, and the mixed solvent coupling procedure might have contributed to the lack of specificity shown by this column material.

A column was then prepared of Sepharose derivative (5), and a similar cell-free extract was applied. Despite a smaller figure of 1.4 μmoles of inhibitor bound per ml of sedimented Sepharose, activity was efficiently absorbed during elution with 0.01 M phosphate buffer. Again, quantitative recovery of purified enzyme was accomplished with 0.2 M phosphate buffer, pH 7. The specific activity of this product was higher, approximately 1.5. However, when aliquots taken from the leading edge, middle and tail of the active protein peak were subjected to polyacrylamide gel electrophoresis, the patterns shown in Fig. 5 were obtained[§].

[§] Similar results were obtained when column materials (3) and (5) were used in attempts to purify the synthetase from T_2 phage-infected *E. coli*.

These results suggested that an ion-exchange effect, caused by the excess unreacted succinyl carboxyl groups, was being superimposed on any specific affinity effect. These groups were, therefore, blocked with glycinamide, and a further sample of cell-free extract was applied to this new column material. The column was made up from 15 ml of the sedimented Sepharose derivative and showed a capacity to absorb 2.32 units of enzyme activity/ml at low phosphate buffer concentrations. Stepwise elevation of the buffer strength to 0.2 M led to >75% recovery of activity. Combining the principal active fractions and re-applying them to the same column using the same stepwise buffer elution, led to the recovery of protein with a specific activity of >3.3. The initial specific activity was 0.0087; overall, a 350-fold purification was achieved with >50% recovery of active protein. The sharpest band of activity was obtained if the final elution step for the second column was performed with 0.3 M phosphate buffer, rather than the 0.2 M buffer used in the first column. Fig. 6 shows the pattern obtained when aliquots of the starting material (A), the product of the first column (B), and the material from the final purification step (C) were subjected to polyacrylamide gel electrophoresis.

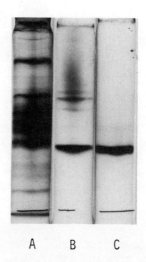

A   B   C

FIG. 6. Electrophoresis of: (A) a crude cellular extract of amethopterin-resistant *L. casei*; (B) the pooled fractions of peak enzymatic activity recovered from the first passage through the variant column material (5) containing glycinamide-blocked carboxyl groups; and (C) the pooled active fractions after a second passage through the same column (cf. legend, Fig. 5, for further details).

SUMMARY

Affinity matrices have been prepared in which fluorodeoxyuridylate, a potent inhibitor of thymidylate synthetase, has been coupled to activated Sepharose by a variety of linking agents. The chemical and physical properties, including K_i values, shown by homogeneous inhibitor derivatives, for the synthetase from both amethopterin-resistant $L.$ $casei$ and the T_2 phage-infected $E.$ $coli$ are reported. It appears that the large molecular weight of the synthetases ($\sim$60,000) requires the inhibitor to be well-extended from the polysaccharide backbone to ensure good interaction. The most effective column material contained a diaminohexane/succinic anhydride/p-aminophenyl spacer attached to the 5'-phosphate group of the inhibitor. In addition, excess succinyl carboxyl groups were blocked with glycinamide. Such a column material showed a high capacity for absorbing enzyme activity (2.32 units/ml) from a crude cell-free extract of $L.$ $casei$ and gave a 350-fold purification of active protein, with a final specific activity of >3.3. The purified enzyme showed a single protein band on polyacrylamide gel electrophoresis.

ACKNOWLEDGMENTS

The authors are indebted to Dr. R.B. Dunlap for his stimulus at the conception of this project, to Dr. J. Galivan for contributing data relating to the T_2 phage-infected $E.$ $coli$ system, and to Drs. N.G.L. Harding and F.M. Huennekens for their advice and encouragement during the course of this work. The work was supported by a research grant from the National Cancer Institute, National Institutes of Health (CA11778).

REFERENCES

1. Wahba, A.J., and Friedkin, M., J. Biol. Chem. 236, PC11 (1961).
2. Pastore, E.J., and Friedkin, M., J. Biol. Chem. 237, 3802 (1962).
3. Reyes, P., and Heidelberger, C., Mol. Pharmacol. 1, 14 (1965).
4. Blakley, R.L., J. Biol. Chem. 238, 2113 (1963).
5. Crusberg, T.C., and Kisliuk, R.L., Fed. Proc. 28, 473 (1969).
6. Dunlap, R.B., Harding, N.G.L., and Huennekens, F.M., Biochemistry 10, 88 (1971).
7. Whiteley, J.M., Jackson, R.C., Mell, G.P., Drais, J.H., and Huennekens, F.M., Arch. Biochem. Biophys. 150, 15 (1972).
8. Dann, J.G., Harding, N.G.L., Newbold, P.C.H., and Whiteley, J.M., Biochem. J. 127, 28P (1972).
9. Leary, R.P., and Kisliuk, R.L., Prep. Biochem. 1, 47 (1971).

10. Cohen, S.S., Flaks, J.G., Barner, H.D., Loeb, M.R., and Lichenstein, J., Proc. Nat. Acad. Sci. (U.S.A.) 44, 1004 (1958).
11. Axen, R., Porath, J., and Ernback, S., Nature (London) 214, 1302 (1967).
12. Danenberg, P.V., Langenback, R.J., and Heidelberger, C., Biochem. Biophys. Res. Commun. 49, 1029 (1972).
13. Whiteley, J.M., Jerkunika, I., and Deits, T., submitted to Biochemistry.
14. Tener, G.M., J. Amer. Chem. Soc. 83, 159 (1961).
15. Cuatrecasas, P., Wilchek, M., and Anfinsen, C.B., Biochemistry 8, 2277 (1969).
16. Chen, H.-C., Craig, L.C., and Stauer, E., Biochemistry 11, 3559 (1972).
17. Galivan, J., Maley, G.F., and Maley, F., Fed. Proc. 32, 2116 (1973).

THE BIOSYNTHESIS OF RIBOFLAVIN: AFFINITY CHROMATOGRAPHY PURIFICATION OF GTP-RING-OPENING ENZYME

L. Preston Mercer and Charles M. Baugh

Department of Biochemistry, University of South Alabama, College of Medicine, Mobile, Alabama

The ability of guanine compounds to serve as precursors of riboflavin has been well established (1,2). This ability was demonstrated in two different organisms in which purine-purine interconversions were blocked to insure retention of radioactive label in the specified molecule. The direct participation of the ribosyl moiety of a guanine nucleoside or nucleotide has not been established. The observation that loss of carbon 8 of guanine precedes riboflavin biosynthesis has led to its comparison with pteridine biosynthesis, which has a similar beginning (3,4). While the ribose moiety of GTP has been shown to be incorporated into pteridines (5) it does not appear to be directly incorporated as the ribityl side chain of riboflavin (6). Therefore, it is not known whether expulsion of carbon 8 of the imidazole portion of GTP represents a common first step in the biosynthesis of both vitamins or the initiation of two independent but similar pathways. (Figure 1.)

It is of interest to study the enzymes which catalyze the release of carbon 8 from guanosine triphosphate and classify them according to products. These enzymes have been studied in cell-free systems (7,3), and partially purified forms (5,8,9).

The use of affinity chromatography for the selective purification of specific enzymes has met with a high degree of success. Robberson and Davis (10) showed that periodate oxidized RNA and UMP could be coupled to

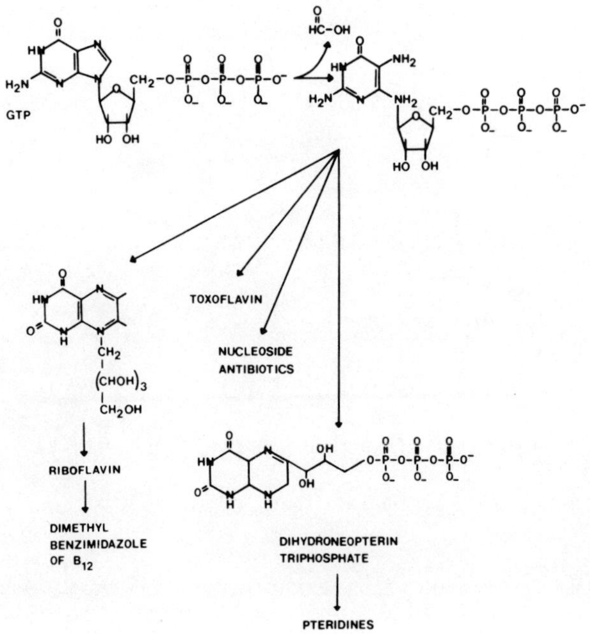

Fig. 1. *Hypothetical tetra-aminopyrimidine intermediate in the biosynthesis of guanine related compounds produced by loss of carbon 8.*

agarose. Their procedure involved the coupling of ε-amino caproic acid methyl ester to cyanogen bromide activated sepharose followed by hydrazinolysis. The oxidized nucleotide was then reacted with the hydrazide to form a stable Schiff's base. This procedure has been adapted to GTP by Jackson and Shiota (9). This paper describes the use of a modification of this technique to prepare an affinity column for GTP ring opening enzymes.

Materials

Cyanogen bromide activated sepharose was purchased from Pharmacia Fine Chemicals Inc., diaminohexane from Aldrich Chemical Co., GTP from Sigma, GTP-8-C^{14} from Amersham Searle Corp. and vitamin-free casein hydrolysate from Nutritional Biochem. Co.

Methods

Bacillus Megaterium, ATCC 10778, a commerical riboflavin producer was chosen as a source of enzyme. The microorganism was grown in a simple salts medium containing 1% vitamin-free casein hydrolysate and 2% glucose. Growth was allowed to proceed for 24 hours at 30°C with forced aeration. The cells were harvested by centrifugation and washed with isotonic saline. They were then ruptured by passage through a French pressure cell and the cellular debris discarded after centrifugation for 30 minutes at 15,000 RPM.

The supernatant was brought to 50% saturation with $(NH_4)_2SO_4$ and the pellet discarded. The remaining protein was collected from the solution by increasing the $(NH_4)_2SO_4$ concentration to saturation. This pellet was resuspended in 0.1 M TES buffer, pH 7.1, and dialyzed overnight vs .001 M EDTA, .001 M TES, pH 7.1.

The dialysate was then poured over a 3 x 3 cm column of DEAE Cl^- equilibrated with the same buffer. The column was washed with buffer and eluted with 1.0 M NaCl in buffer. The enzyme containing effluent was then dialyzed as above in preparation for affinity chromatography.

The affinity column was prepared from cyanogen bromide activated sepharose. The sepharose was washed with H_2O and brought to pH 9.8 with 0.1 N $NaHCO_3$. Diaminohexane was dissolved in water and adjusted to pH 10 with concentrated HCl. Using a ratio of 2 mMoles of amine to 1.0 ml packed gel, the mixture was rotated in the cold room overnight (11). The gel was then washed to prepare for coupling with GTP. (Figure 2.)

GTP was mixed with $NaIO_4$ on a 1:1 molar ratio in citrate-PO_4 buffer pH 3. The mixture was incubated at room temperature for 30 minutes and added to 5 ml of the gel preparation. The resulting suspension was stirred at room temperature for 2 hours after which the gel was filtered and washed. The dialysate from the DEAE column was then brought to 0.30 M with respect to NaCl and poured through a 1 x 3 cm column of the affinity gel. The gel was washed with 0.30 NaCl in .01 M TES pH 7.1 and subsequently eluted with 1.0 M NaCl in the same buffer.

At this point, the enzyme was contained in about 5-10 ml 1.0 M NaCl in buffer. Desalting and concentration were carried out simultaneously using an Amicon stirred cell and a UM20E ultrafiltration membrane. The volume was reduced to 1.0 ml in one hour at 50 lbs./sq. in. nitrogen pressure.

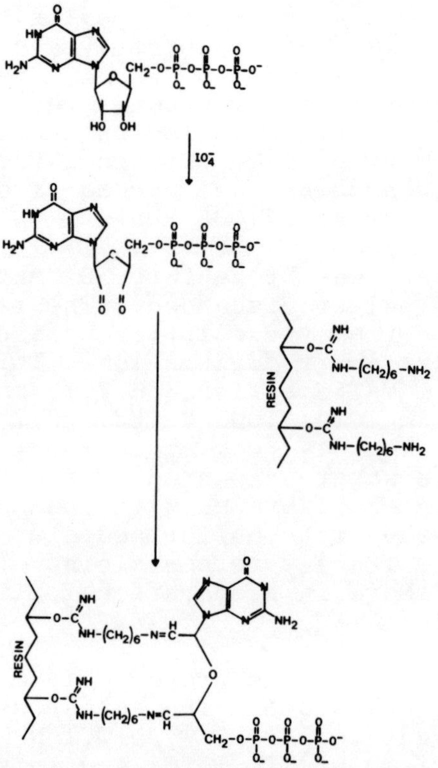

Fig. 2. Outline of the production of the dialdehyde analog of GTP by periodate oxidation and its attachment to diaminohexane substituted sepharose.

Enzyme activity was monitored using the ethyl acetate method described by Cone and Guroff (8). An incubation mixture contained 0.5 ml 0.1 M TES pH 7.85, 100 mµCi GTP-8-C^{14}, (30 mCi/mMole), 3 mM $MgCl_2$, and enzyme to a total of 0.65 ml. The mixture was incubated for one hour at 30° and then terminated with 2 ml 0.1 N HCl. The radioactive formate released enzymatically was then extracted with 2 ml ethyl acetate, plated in Multisol and counted. Controls were run using protein denatured by prior addition of 5% TCA. The protein content was determined by the method of Lowry (12).

Results

To determine the proper ligand to bind to the resin, the crude protein solution was assayed for ability to release formate from guanine-8-C^{14}, guanosine-8-C^{14}, and GMP, GDP, and GTP-8-C^{14}. Activity was obtained only with GTP, therefore, this compound was employed for specific binding to sepharose. The method developed produced substituted sepharose containing 3.9 μmoles of GTP per ml of settled gel. This figure was determined by coupling GTP-8-C^{14} of known specific activity to the gel. After coupling, an aliquot of gel was counted in 4% carbosil in Multisol and coupling efficiently calculated.

Binding of the desired protein to the GTP-substituted dextran was incomplete due to interference by non-specific protein adsorption. Addition of crude enzyme to the affinity column showed incomplete binding. In a typical experiment, 5 ml crude protein was added to the gel and allowed to run through. The effluent was collected and the gel washed with five 5 ml portions of 0.1 M buffer. The gel bed was then stripped with three 5 ml portions of 1.0 M buffer and the fractions assayed for activity. (Table I)

Table I

Binding of GTP-8-C^{14} ring opening activity to substituted sepharose

5 ml crude protein was applied to the gel and the effluent collected. The column was washed with successive 5 ml aliquots of 0.1 M buffer until no more activity was eluted. The column was then eluted with 1.0 M buffer until all activity was accounted for. Binding was incomplete as noted by activity in 3 and 4.

Buffer-5 ml/aliquot	Activity/.05 ml effluent
1. Addition of protein	0
2. 0.1 M buffer	0
3. 0.1 M buffer	5632
4. 0.1 M buffer	3187
5. 0.1 M buffer	98
6. 0.1 M buffer	0
7. 1.0 M buffer	4770
8. 1.0 M buffer	4970
9. 1.0 M buffer	77

Binding was incomplete even though theoretical considerations implied a capacity sufficient for the activity applied; therefore, a gradient elution of the gel was undertaken using a linear sodium chloride gradient of zero to 1.0 M, total volume 100 ml. The results showed elution of inactive protein, thus indicating interference by nonspecific interactions. (Figure 3.)

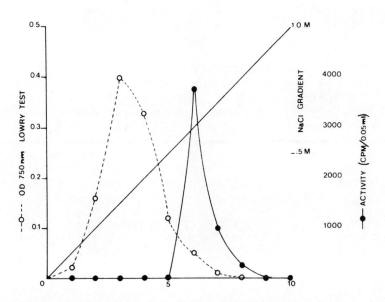

Fig. 3. Elution pattern of protein from substituted sepharose. Protein concentration indicated by --O-- and enzyme activity indicated by —●—. Gradient was linear 0 → 1 M NaCl total volume 100 ml, 10 ml/fraction.

Subsequent experiments were designed to determine the salt concentration which would minimize nonspecific adsorption of the protein applied to the column without interference with affinity binding. When the crude solution was made 0.3 M in NaCl before application to the column, little nonspecific binding was noted while all ring opening activity was bound. This technique greatly enhanced purification. The overall purification was 880 fold with a yield of 70%. These data are summarized in Table II.

The enzyme has been only partially characterized at this time. Stability is sufficient to allow an accumulation of enzyme so that mechanistic studies can be readily carried out. The properties already determined are reported in Table III.

Table II

Overall purification of GTP ring opening activity

	Total Act.	Specific Act.
I. Crude	3,330,000	4,666 cpm/mg
II. 50-100% Fraction	3,960,000	10,714 cpm/mg
III. DEAE Eluate	2,900,000	15,710 cpm/mg
IV. Affinity Column Eluate	2,200,000	4,105,000 cpm/mg

Table III

Properties of GTP ring opening enzyme

Enzyme was much more stable if EDTA ($10^{-3}M$) was present from the first step. Inhibition by phosphate was complete at 0.1 M while Tris and dithioerythritol were only partially inhibitory at this concentration. An absolute requirement for Mg^{++} was not shown although enhancement of activity was noted up to 3 mM. The same stimulation of activity required 30 mM NH_4^+.

pH Max: 7.85 at 30°C
Inhibited by: -SH reagents, Tris, $HPO_4^=$
Stimulated by: Mg^{++}, NH_4^+
Stabilized by: EDTA

Discussion

The use of affinity chromatography has been of great value in the rapid, specific purification of a variety of molecules. As an adjunct to classical protein purification it often leads to increased purity with greater recovery.

In the described procedure, cyanogen bromide activated sepharose was reacted first with diaminohexane to provide a spacer between the specific ligand and the gel matrix. The GTP was then attached through a Schiff's base linkage which showed good stability over long periods of time. This resulted in a resin bed with high flow characteristics and good substitution. The GTP analog retained the structural characteristics necessary for enzyme recognition, and no degradation of substrate could be detected. This lack of enzyme activity can be rationalized in several ways. Conditions were used in which the optimum pH and temperature were not present. Also, while recognition of substrate obviously took place, the structure may have been modified sufficiently to prevent catalysis.

The classical method of elution employed in affinity chromatography is the use of substrate or substrate analog containing buffer to displace protein attached to gel bound ligand leaving behind the non-specifically bound protein. However, an attractive alternative to this is the use of gradient elution with salt. As noted, this can result in purification due to differential elution of non-specifically bound protein. Also, in many cases, a ligand may offer a binding site for more than one protein. Under these conditions, gradient elution may offer a method of differential separation based on different binding affinities. Thus a combination of techniques may offer greater control of purification. A third alternative is the experimental determination of an ionic strength which eliminates non-specific interaction of protein with dextran. This allows one to apply the desired enzyme in a much cruder form, thereby reducing its handling. The results are more rapid procedures leading to greater recovery of activity.

Bibliography

1. Baugh, C.M. and C.L. Krumdieck, *J. Bacteriol.*, **98**: 1114 (1969).

2. Bacher, A. and F. Lingens, *Angew. Chem. Internat. Edit.*, 8:371 (1969).
3. Shiota, T. and M.P. Palumbo, *J. Biol. Chem.*, 240: 4449 (1965).
4. Plaut, G.W.E., *J. Biol. Chem.*, 211:111 (1954).
5. Burg, A.W. and G.M. Brown, *J. Biol. Chem.*, 243:2349 (1968).
6. Baugh, C.M. and L.P. Mercer, unpublished results.
7. Reynolds, J.J. and G.M. Brown, *J. Biol. Chem.*, 239: 317 (1964).
8. Cone, Joyce and Gordon Guroff, *J. Biol. Chem.*, 246: 979 (1971).
9. Jackson, R. and T. Shiota, *Biochim. Biophys. Res. Comm.*, 51:449 (1973).
10. Robberson, D.L. and N. Davidson, *Biochem.*, 11:533 (1972).
11. Cuatrecasas, Pedro, *J. Biol. Chem.*, 245:3059 (1970).
12. Lowry, O.H., et al, *J. Biol. Chem.*, 193:265 (1951).

PURIFICATION OF TYROSINE-SENSITIVE 3-DEOXY-D-ARABINOHEPTULOSONATE-7-PHOSPHATE (DAHP) AND TYROSYL-tRNA SYNTHETASES ON AGAROSE CARRYING CARBOXYL-LINKED TYROSINE

Andrew R. Gallopo, Department of Chemistry, Montclair State College, Upper Montclair, N.J. 07043

Philip S. Kotsiopoulos and Scott C. Mohr*, Department of Chemistry, Boston University, Boston, Ma. 02215

In principle any ligand-binding site with an appreciable degree of specificity can be used to purify a macromolecule by affinity chromatography. We have prepared a resin which contains covalently bound tyrosyl groups and employed it to purify one enzyme (tyrosyl-tRNA synthetase from E. coli) which interacts with the resin via its substrate-binding site and another (DAHP synthetase from yeast) which interacts via an effector-binding site. Our resin which contains a hexamethylene diamine (HMD) extension between the agarose matrix and the ligand appears to bind DAHP synthetase much more strongly than the previously used "tyrosine-Sepharose" which lacked such a spacer (1).

Initially we intended to prepare a resin which would be effective in the purification of the tyrosyl-tRNA synthetase. This consideration dictated the mode of attachment of the tyrosine since kinetic studies (2, 3) indicate that binding of tyrosine to the active site of this enzyme requires a free (protonated) α-amino group as well as the phenolic side chain. Various substitutions of the carboxylate moiety have little effect on the binding. With this in mind we set about to make a resin with the structure given in Figure 1.

$$HO-\underset{}{\bigcirc}-CH_2-\underset{\underset{NH_3^{\oplus}}{|}}{CH}-\overset{O}{\underset{||}{C}}-\underset{H}{N}-(CH_2)_6-\underset{H}{N}-\{ \text{(agarose)}$$

Figure 1. <u>Carboxyl-linked Tyrosyl Agarose</u>
with an HMD Spacer

* To whom correspondence should be addressed.

The following equations outline our synthesis procedure:

A. L-tyrosine + carbobenzoxychloride $\xrightarrow[0°C]{2N\ NaOH}$ N,O-diCBZ-L-tyrosine

N,O-diCBZ-L-tyrosine $\xrightarrow[\substack{anhyd.\ C_6H_6 \\ 0°C}]{PCl_5}$ O-CBZ-L-tyrosine-N-carboxyanhydride

O-CBZ-L-tyrosine-NCA $\xrightarrow[\substack{Et_2O/EtOAc \\ 0°C}]{HBr(g)}$ L-tyrosine-NCA

B. agarose (Sepharose 4B) $\xrightarrow[\substack{pH\ 11 \\ 20°C}]{BrCN}$ 'activated' agarose

'activated' agarose + $H_2N-(CH_2)_6-NH_2$ $\xrightarrow[\substack{pH\ 10 \\ 0°C}]{}$ HMD-agarose

C. HMD-agarose + L-tyrosine-NCA $\xrightarrow[\substack{pH\ 10.5 \\ 4°C}]{DMF/H_2O}$ L-tyrosyl-HMD-agarose

Part A, the preparation of L-tyrosine-N-carboxyanhydride represents a slight modification of the procedure of Katchalski and Sela (4). We prepared HMD-agarose (part B) according to the method of Cuatrecasas (5).

The final product gave a positive ninhydrin reaction and a positive Pauly test for tyrosine. We have not carried out a quantitative determination of the concentration of tyrosyl groups, but based on reasonable yields in steps B and C above, we estimate 5-10 μmole tyrosine bound per milliliter of agarose gel.

Figure 2 shows the result of applying our resin to the purification of tyrosyl-tRNA synthetase from E. coli. The enzyme emerged midway in the broad protein peak eluted from the column by a 0.1-0.5M NH_4Cl gradient. Clearly some affinity exists between the enzyme and the resin, but it remains to be determined whether the effect reflects specific ligand-binding or merely ion-exchange processes. The tyrosyl-agarose column effected a 3.5-fold purification of the enzyme under these conditions. Further work will be necessary to establish (a) whether or not this is true affinity

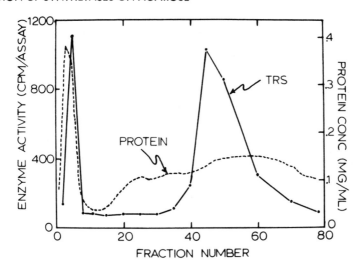

Figure 2. Elution pattern of tyrosyl-tRNA synthetase (TRS) chromatographed on carboxyl-linked tyrosyl agarose. 10.5 ml of peak fractions from the first DEAE cellulose column of Calendar and Berg (6) containing approximately 25 mg total protein in 20% glycerol was applied to a 1 x 8 cm column of tyrosyl agarose which had been equilibrated with 0.01M sodium phosphate, pH 6.7, 1 mM 2-mercaptoethanol, 0.1 mM EDTA. The column was washed with 30 ml of this buffer and then eluted with a linear gradient of 0-0.5M NH$_4$Cl in the same buffer (total volume of gradient: 300 ml). The column flowed at a rate of 1.0 ml/min and 4-ml fractions were collected. Protein was measured by the Lowry method and TRS activity by the charging assay using ^{14}C-tyr. (7)

chromatography and (b) whether altered conditions (buffer, ionic strength, pH) will permit a higher purification factor.

Use of affinity chromatography to separate the phenylalanine- and tyrosine-sensitive isozymes of yeast 3-deoxy-D-arabino-heptulosonate-7-phosphate (DAHP) synthetase has already been reported (1). The resin chosen for this purpose ("tyrosine Sepharose") consisted of tyrosine directly coupled to activated agarose through the amino group. While successful separation of the two isozymes did occur, the resin only retarded the tyrosine-sensitive DAHP synthetase, but did not adsorb it strongly.

Moreover, the two isozymes invariably overlapped in the elution pattern. Reviewing this work, Cuatrecasas (8) has remarked "it is tempting to speculate that strong adsorption of this enzyme would occur if the tyrosyl group were placed farther from the matrix backbone."

Since our resin fulfilled this requirement - albeit with a different ionizable group available on the tyrosine moiety - we decided to test it with the yeast DAHP synthetase system. Figure 3 shows the results. (Note that the figure is presented in such a way as to emphasize the purification effect - the ordinate gives enzyme specific activity not total enzyme activity.)

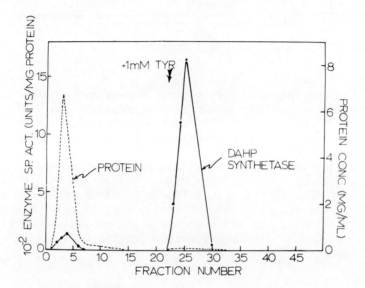

Figure 3. Elution pattern of DAHP synthetase chromatographed on carboxyl-linked tyrosyl agarose. 6.5 ml of dialyzed protamine sulfate supernatant (9) containing approximately 130 mg total protein was applied to a 0.75 x 12 cm column of tyrosyl agarose which had been equilibrated with 0.2M sodium phosphate, pH 6.5, 0.1 mM $CoSO_4$, 0.1 mM phenylmethanesulfonyl fluoride. The column was eluted with the same buffer until the protein concentration in the effluent dropped to virtually zero. At that point (arrow) 1 mM L-tyrosine was added to the buffer and elution continued. The column flowed at a rate of 0.17 ml/min and 5-ml fractions were collected. Protein was measured by the Lowry method and DAHP synthetase essentially according to Takahashi and Chan (9)(see also Table I). Enzyme assays in the breakthrough peak (fractions 1-6) were performed in the presence of 1 mM tyrosine and those for the peak of adsorbed enzyme (fractions 23-27) were performed in the presence of 2 mM phenylalanine.

Clearly our carboxyl-linked tyrosyl agarose with an HMD spacer effects a superior separation of the DAHP synthetase isozymes. The affinity column gives a 280-fold purification of the tyrosine-sensitive isozyme in a single step (Table I). We have also repeated the control experiment (1) with unsubstituted agarose and confirmed that this material retards neither isozyme. All our experiments have been done on a small scale, but scale-up should be quite straight-forward. The affinity resin can be easily regenerated by treatment with 0.5M sodium phosphate buffer followed by equilibration with starting buffer.

Table I. Purification of Yeast DAHP Synthetase Isozymes

Fraction	Volume (ml)	Protein conc. (mg/ml)	Protein (mg)	Specific activity[a] (units/mg protein)
1. Centrifuged extract	150	330(?)	49,500	0.7
2. Protamine sulfate supernatant	125	65	8100	3.7
3. Dialyzed p.s.s.	130	20	2600	12.6
4. Affinity[b,c] chromatography				
Peak I (fractions 2-6)	300	4.73	1400	333
Peak II (fractions 23-27)	300	0.07	21	3570

[a.] One unit of enzyme catalyzes the formation of one nanomole of DAHP/min/ml reaction mixture.

[b.] Figures in the table represent the amounts which would be expected if the affinity column were scaled-up to accept the whole preparation.

[c.] Appropriate assays in the presence and absence of phenylalanine and tyrosine demonstrate that peak I contains essentially the phe-sensitive isozyme and peak II the tyr-sensitive isozyme.

Comparing our results with those of Takahashi and Chan (9) we were struck by the anomaly that affinity resins containing amino-linked or carboxyl-linked tyrosine both appear to adsorb the yeast tyrosine-sensitive DAHP synthetase. Whether the greater affinity exhibited by our resin stems solely from the effect of the HMD spacer or can be partly attributed to the presence of the free α-amino group is not clear. Kinetic data (9) showing inhibition of this isozyme by glycyl-L-tyrosine but not by L-tyrosinamide have been interpreted to mean that the free carboxyl group, but not a free amino group is required for binding. On the other hand, L-phenylalanylglycine inhibits the isozyme to some extent (9) and in preliminary experiments we have found that L-tyrosine methyl ester also acts as an inhibitor*. One obvious conclusion is that the enzyme might interact almost exclusively with the phenolic side chain of tyrosine and be largely indifferent to the state of the amino and carboxyl groups.

Carboxyl-linked tyrosyl-HMD-agarose should find numerous applications in enzymology. We anticipate that systematic study of the conditions of chromatography will lead to its more effective use in the purification of tyrosyl-tRNA synthetase. An extensive report (10) on the purification of methionyl-tRNA synthetase on methionyl-HMD-agarose appeared while this work was in progress and indicates some of the parameters to investigate.

ACKNOWLEDGEMENTS

We thank Ms. Sylvia Betcher-Lange for assistance with the TRS purification and Dr. Richard A. Laursen for helpful discussions as well as the donation of a sample of L-tyrosine methyl ester. This research was supported in part by a grant (no. E-618) from the American Cancer Society.

REFERENCES

(1) Chan, W. W.-C. and Takahashi, M. (1969) Biochem. Biophys. Res. Commun. 37, 272

(2) Calendar, R.C. and Berg, P. (1966) Biochemistry 5, 1690

(3) Santi, D.V. and Peña, V.A. (1972), personal communication.

(4) Katchalski, E. and Sela, M. (1953) J. Am. Chem. Soc. 75, 5288.

(5) Cuatrecasas, P. (1970) J. Biol. Chem. 245, 3059.

---

* 1 mM L-tyrosine methyl ester gave 50% inhibition.

(6) Calendar, R.C. and Berg, P. (1966) Biochemistry 5, 1681.

(7) Nishimura, S. and Novelli, G.D. (1964) Biochim. Biophys. Acta 80, 574.

(8) Cuatrecasas, P. (1972) Adv. Enzymol. 36, 29.

(9) Takahashi, M. and Chan, W. W.-C. (1971) Can. J. Biochem. 49, 1015

(10) Robert-Gero, M. and Waller, J.-P. (1972) Eur. J. Biochem. 31, 315.

STRUCTURAL REQUIREMENT OF LIGANDS FOR AFFINITY CHROMATOGRAPHY ABSORBENTS: PURIFICATION OF ALDEHYDE AND XANTHINE OXIDASES

Albert E. Chu	Sterling Chaykin
Bio-Rad Laboratories	Dept. of Biochemistry & Biophysics
Richmond, California	Univ. of California, Davis, Calif.

Liver aldehyde oxidase catalyzes the oxidation of N-methyl nicotinamide to a mixture of N-methyl-4-pyridone-3-carboxamide and N-methyl-2-pyridone-5-carboxamide (1). In the past this enzyme has been purified from the livers of a number of different mammals using classical salt, organic solvent, gel and ion-exchange procedures (2). These methods suffered from being inefficient in terms of both enzyme yield and the time required to conduct purification (3). Perhaps more importantly, there is reason to believe that the homogenous enzymes so isolated were contaminated with inactive enzyme (3). In this paper, a procedure for the isolation of aldehyde oxidase by affinity chromatography will be presented. Using this method, the enzyme is recovered in high yield and with a maximum specific activity equal to or greater than that obtainable by previous methods. The procedure requires six hours to carry out and produces, as a fringe benefit, a highly purified xanthine oxidase as well.

The first observation which suggested that affinity methods might be employed in the isolation of aldehyde oxidase is shown in Table 1. It is the 30-fold lesser K_m exhibited by N-benzyl nicotinamide as compared to the normal substrate N-methyl nicotinamide (3). Of all the nicotinamide derivatives thus far studied, N-benzyl nicotinamide has the lowest K_m (4). Since K_m can have both binding and kinetic implications the relative binding strengths of methyl and benzyl substituents were studied using non-substrates. Substrate itself could not be used as a ligand because H_2O_2 produced by enzyme action on a substrate ligand would inactivate the enzyme. The remaining data in this table show that binding of N-substituted 6-methyl

TABLE I

Substrates	K_m
n-methyl nicotinamide	3×10^{-4} M
n-benzyl nicotinamide	1×10^{-5} M

INHIBITORS Derivatives of nicotinamide	K_i M^{-1}	Type of inhibition
N-methyl-6-methyl	7×10^{-5}	Noncompetitive
N-benzyl-6-methyl	3×10^{-5}	Competitive

FIGURE 1

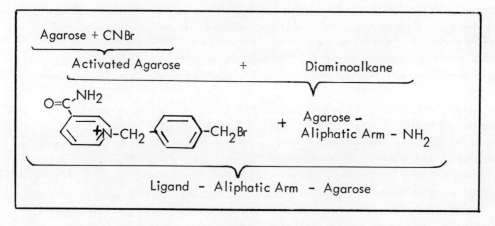

nicotinamide inhibitors is increased when a benzyl group is substituted for methyl. On the basis of these data, a first affinity column was constructed as shown on Figure 1.

Agarose was activated by reaction with cyanogen bromide, then reacted with an aliphatic diamine arm and the ligand, N-(4-bromo methyl)-benzyl nicotinamide (4). When a sample of a heat-treated preparation (after heat step in the method described in reference 5) of a rabbit liver homogenate was applied to a column constructed of this affinity matrix, aldehyde oxidase was retained. The majority of the protein flowed through the column. A sodium chloride gradient eluted the bound oxidase. Although a 10-12 fold purification of the enzyme was achieved, aldehyde oxidase was not separated from xanthine oxidase. Alteration of the elution conditions did not improve the extent of the purification. Increasing the length of the aliphatic arm to 25 atoms, with 22 carbon and 3 nitrogen atoms, led to a somewhat greater affinity of the enzyme for the matrix. However, it was not until a 6-methyl substituted N-benzyl nicotinamide ligand was used that it was clear that an unusually good affinity matrix had been discovered. Although salt could cause most of the protein to be eluted from a matrix constructed using the 6-methyl nicotinamide ligand the enzyme was not eluted until the protonated nitrogen atoms in the aliphatic arm had been neutralized. When a glycine-sodium hydroxide gradient was used, the enzyme was eluted between pH 10 and 11.0. Again, it was contaminated with xanthine oxidase. Subsequently it was found that the use of a glycine-sodium hydroxide gradient in the presence of 0.5 M NaCl led to a separation of xanthine oxidase and aldehyde oxidase with over 100-fold purification of aldehyde oxidase in 80% yield. A typical elution pattern is shown in Figure II.

The peak on the left represents the bulk of the protein. It flowed through the column without retention. The middle peak, xanthine oxidase, eluted between pH values of 8.0 and 9.5. Aldehyde oxidase is shown on the far right of this figure. It eluted between pH values of 9.8 and 10.5. The specific activity (1.6 umole/ml/min) of the enzyme after affinity chromatography under these conditions was greater than that of the homogenous enzyme isolated by previous procedures (3). Electrophoretic methods showed the rabbit liver aldehyde oxidase to be slightly more than 50% pure. Thus, the homogenous enzyme from rabbit liver should have a specific activity greater than 3.0 and earlier preparations of the homogenous enzyme must have contained inactive molecules. The potential applicability of this affinity method for the isolation of aldehyde oxidase from other sources is emphasized by the fact that the method has been successfully used for the isolation of the enzyme from hog liver.

FIGURE II

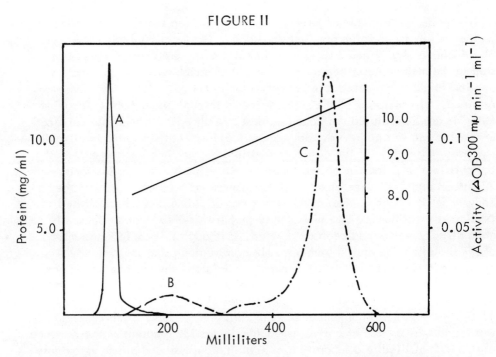

Purification of rabbit liver aldehyde oxidase on an affinity column of the following ligand matrix:

Agarose-NH-C_{10}--$\overset{+}{NH_2}$-C_2-$\overset{+}{NH_2}$--C_{10}--$\overset{+}{NH_2}$-CH_2-⟨⟩-CH_2-$\overset{+}{N}$⟨⟩$\overset{H_2N}{\underset{H_3C}{}}$C=O

A rabbit liver heat treated sample (244 mg) was applied to the column. A linear gradient of decreasing glycine concentration and increasing pH and sodium chloride concentration was used. The mixing chamber contained 0.05 M glycine, the companion chamber contained a solution which has 0.05 M KOH and 1 M NaCl.

Peak A is protein in the flow-through peak.
Peak B is xanthine oxidase activity.
Peak C is aldehyde oxidase activity.

The unusual characteristics of the purification of aldehyde oxidase using the 25 atom aliphatic arm did not appear to be solely related to its length. The positively charged amino groups in the arm also appeared to be important. Several aliphatic arm analogues were synthesized using N-benzyl-6-methyl nicotinamide as ligand. The results are shown in Figure III.

In general, 80% recovery of the enzyme activity was achieved. However, the extent of enzyme purification was greatest when the ligand was 11 atoms removed from the first positively charged amino group in the arm. When there was no charge on the arm, the special binding properties of the column and its potential for extensive purification were lost. Given the crucial nature of the disposition of a positively charged group, the length of the arm was not of paramount importance. Arm lengths of 11 and 25 atoms were both good although the 25 atom arm was best. An additional benefit of the 25 atom arm was that only in this case were xanthine oxidase and aldehyde oxidase fully separated. It would therefore appear that the inclusion of ionic components in the aliphatic side chain can result in increased power of the affinity chromatography method. This device would be particularly useful in cases where ligands of marginal affinity must be employed, that is, with binding constants on the order of 10^{-5} to 10^{-6}.

FIGURE III

Arm	Arm Length (Atoms)	Percentage of Oxidase Recovery	Maximum Fold Purification
$A-O-(Affi-Gel\ 10)-\overset{O}{\overset{\|}{C}}-NH-C_{10}-L$	20	100	~ 10
$A-NH---C_{10}---L$	11	80	30
$A-NH---C_9-\overset{+}{N}H_2-C_2-\overset{+}{N}H_2-C_9-L$	23	80	20
$A-NH--C_{10}--\overset{+}{N}H_2-C_2-\overset{+}{N}H_2-C_{10}-L$	25	75	>100
$A-NH--C_{12}--\overset{+}{N}H_2-C_2-\overset{+}{N}H_2-C_{12}-L$	29	80	16

A = Agarose   (Affi-Gel 10) = $A-O-CH_2CH_2CH_2-NH\overset{O}{\overset{\|}{C}}-CH_2CH_2-$

$L = -\overset{+}{N}H_2-CH_2-\langle\text{phenyl}\rangle-CH_2-\overset{+}{N}\langle\text{pyridyl with } NH_2\text{-CO, } CH_3\rangle$

Conclusions:

1. An affinity column has been described which has the power to permit the easy isolation of highly purified aldehyde oxidase from liver homogenates in good yield.

2. This method also permits the simultaneous isolation of a highly purified xanthane oxidase.

3. A new parameter, namely, the introduction into the aliphatic side arm of ionic groups correctly disposed with respect to the affinity ligand, has been found to increase the power of affinity chromatography.

References:

1-a. Felsted, R. L. and Chaykin, S., J. Biol. Chem. 242, 1274 (1967)

  b. Murashije, J., McDaniel, D. and Chaykin, S., Biochem. Biophys. 118, 556 (1966)

  c. Huff, S. D. and Chaykin, S., J. Biol. Chem. 242, 1265 (1967)

  d. Chaykin, S., Dagani, K., Johnson, L. and Samli, M., J. Biol. Chem. 240, 932 (1965)

2. Rajagopalan, K. V., Fridovich, T. and Handler, P., J. Biol. Chem. 237, 922 (1962)

3. Felsted, R., Ph.D. dissertation 1972

4. Chu, A. E. Y., Ph.D. dissertation 1973

5. Felsted, R. L., Chu, A. E. Y., and Chaykin, S., J. Biol. Chem. 248, 2580 (1973)

PART II

IMMOBILIZED BIOCHEMICALS

IMMOBILIZED POLYNUCLEOTIDES AND NUCLEIC ACIDS

P. T. Gilham

Department of Biological Sciences

Purdue University, West Lafayette, Indiana   47907

The use of immobilized nucleic acids and polynucleotides in the study of nucleic acids and their associated enzymes has become widespread in recent years. There are two main areas of application: the fractionation of nucleic acids and polynucleotides through base-paired complex formation and the isolation and purification of nucleic acid-associated enzymes by affinity chromatography. There are now a number of methods for the preparation of immobilized polynucleotides and the potential uses of these materials depend to some extent on the means by which the polymers are attached to the insoluble supports.

METHODS OF IMMOBILIZATION

As indicated in the following table, the methods for the preparation of immobilized polynucleotides can be broadly divided into two categories: those involving covalent linkages between the polymer and the support, and those employing physical entrapment of the polymer within the support matrix.

In those cases where covalent binding of the molecule is desirable, terminal attachment can be readily achieved by activation of the phosphomonoester group at the polynucleotide terminal in the presence of an insoluble polysaccharide. This reaction yields a product in which the polynucleotide is connected to the support by a stable phosphodiester linkage. A second approach exploits the reactivity of the terminal dialdehyde group resulting from the periodate oxidation of ribopolynucleotides. Condensation of these dialdehyde functions with supports containing primary amine or hydrazide

Table I

1. Covalent Binding at the Polynucleotide Terminals.

    (a) Activation of terminal phosphate group.
    (b) Periodate oxidation of 3'-terminals in RNA species.
    (c) Condensation of RNA 3'-terminals with supports containing dihydroxyboryl groups.

2. Covalent Binding at Multiple Points.

    (a) Condensation with phosphocellulose.
    (b) Reaction with CNBr-activated supports.
    (c) UV-irradiation.

3. Physical Entrapment.

groups allows covalent attachment at the polynucleotide 3'-terminals. A third method arises from the discovery that the terminal cis diol groups in ribopolynucleotides are capable of forming specific cyclic complexes with supports containing covalently-bound dihydroxyboryl groups. The use of phosphocellulose and CNBr-activated supports permits the multiple point attachment of polynucleotides and, in the case of CNBr activation, the reaction mechanism appears to be similar to that operating in the immobilization of enzymes by this method. Immobilization induced by ultraviolet irradiation of polynucleotides probably results in multiple point attachment also, although the mechanism of the reaction is obscure.

Covalent Attachment at Polynucleotide Terminals

Insoluble supports containing small homopolynucleotides can be prepared by the chemical polymerization of the appropriate mononucleotide (1,2). For example, the treatment of thymidine 5'-phosphate with dicyclohexylcarbodiimide in anhydrous solution produces thymidine oligonucleotides of the form, pdT-dT$_n$-dT (n < 20). In the reaction mixture, the terminal phosphate groups are present in an activated form and the addition of dry cellulose to the mixture results in the condensation of these groups with the hydroxyl functions of the support. In this case the oligonucleotides are bound to the support through stable phosphodiester linkages at their 5'-terminals. In an analogous way the polymerization of a nucleoside 3'-phosphate yields immobilized oligonucleotides connected to the cellulose at their 3'-terminals.

Preformed polynucleotides from either synthetic or natural sources may also be bound to cellulose at one of their terminals (3). In this case, the reaction is carried out in aqueous solution and a water-soluble activating agent, N-cyclohexyl-N'-β-(4-methylmorpholinium)ethylcarbodiimide, is used. The carbodiimide activates the terminal phosphomonoester group resulting in, again, a phosphodiester linkage to the support. The method was initially tested with mononucleotides, polynucleotides and tRNA (3), and it has recently been shown that, in the case of immobilized tRNA, practically all of the bound nucleotide material can be released by exposure to pancreatic ribonuclease (4), an observation that confirms the proposed terminal linkage to the support.

The reactivity of the 3'-terminal cis diol group in polyribonucleotides and RNA can be exploited in two ways for the terminal binding of these molecules. Periodate oxidation of the diol moiety to the corresponding dialdehyde yields a product that can undergo condensation with aminoethylcellulose, and the resulting linkage can then be stabilized by reduction with sodium borohydride (5). The binding in these products is thought to arise from the formation of a substituted morpholine ring structure that includes the nitrogen of the aminoethylcellulose and the five atoms that originally constituted the ribose ring of the terminal nucleoside. The assignment of this linkage is based on the structure of the product obtained from the borohydride reduction of the complex formed between periodate-oxidized adenosine 5'-phosphate and methylamine (6). Supports containing hydrazide groups will react in a similar manner with periodate-oxidized polyribonucleotides to form reasonably stable linkages. For example, a mixture of agar and polyacrylic acid hydrazide has been used to immobilize tRNA in this way (7), and reduction with sodium borohydride has also been used to stabilize the linkage in a similar complex obtained from periodate-oxidized tRNA and hydrazinyl-Sepharose (8).

Another method for the terminal binding of polyribonucleotides arises out of the observation that polymers that possess terminal cis diol groups are capable of forming specific complexes with supports containing covalently-bound dihydroxyboryl groups (9-11). N-[N'-(m-Dihydroxyborylphenyl)succinamyl]aminoethylcellulose (9) prepared from the condensation of N-(m-dihydroxyborylphenyl)-succinamic acid with aminoethylcellulose forms cyclic boronate structures at pH 8-9 with molecules containing the ribonucleoside diol group. In certain applications immobilized polynucleotides prepared by this method may have some advantages over those mentioned above in that, at pH 6, the complex breaks down, allowing the recovery of the bound polynucleotide from the support and the substituted cellulose is then available for the binding of another polymer.

Covalent Attachment at Multiple Points

Some of the early attempts to immobilize nucleic acids and polynucleotides exploited an activated form of acetylated phosphocellulose (12). Activation was achieved by treating the derivatized cellulose with dicyclohexylcarbodiimide in a non-aqueous solvent and it is presumed that the subsequent binding of the polynucleotide resulted from the formation of phosphodiester linkages between the support and the various hydroxyl groups on the polynucleotide chain. Thus, it was possible to bind the phosphocellulose to the glucosyl hydroxyl groups of phage T4 DNA (12) and to the 2'-hydroxyl groups of various polynucleotides and RNA molecules (13).

More recently, the immobilization technique that has been used so successfully for protein molecules and other ligands has been applied to polynucleotides. Single stranded RNA and DNA may be efficiently bound at pH 8 to cyanogen bromide-activated agarose (14) and, while the mechanism of the immobilization is not known, it seems likely that binding results from the formation of covalent linkages consisting of isourea ether groups connecting the oxygen atoms of the polysaccharide hydroxyl groups to the nitrogen atoms of some of the nucleotide bases within the polynucleotide chains. It has been suggested that, if the condensation reaction with the activated polysaccharide is carried out at pH 6, polynucleotides may be immobilized by single point attachment at the terminal phosphate groups of the polynucleotide strands (15). However, in this study, control experiments involving polynucleotides that do not possess terminal phosphate groups were not reported, and a more recent investigation of the reaction with cyanogen bromide-activated agarose indicates that multi-point attachment of polynucleotides occurs also at pH 6 (16).

Denatured DNA, RNA and polynucleotides are immobilized when exposed to UV-irradiation in the presence of inert supports such as polyvinyl beads, cellulose, and nylon threads (17). In this procedure it is possible that no covalent linkages are formed between the support and the polynucleotide strands since, in the absence of the support, these polymers are capable of forming insoluble gels upon irradiation. These gels are presumably a consequence of the production of intermolecular cross-links resulting from pyrimidine-pyrimidine dimer formation and the supports may serve only to immobilize the insoluble products so formed. Alternatively, the induction of both inter- and intramolecular cross-links of this type may be responsible for the formation of macrocyclic structures resulting in the physical entrapment of the polynucleotide strands within the support matrix. Binding could also result from UV-catalyzed addition reactions involving functional groups on the supports and the 5,6 double bonds in the pyrimidine moieties of the polynucleotide chains. Nevertheless, the method seems to have some general applicability in that, by exposure to UV-irradiation, calf thymus DNA (18), viral RNA (19), and ribosomal RNA (20) have been bound to cellulose, and poly(U) has been immobilized on fiberglass filters (21).

Physical Entrapment

A number of methods for the physical immobilization of nucleic acids have been devised. With supports such as cellulose acetate (22), agar (22), and polyacrylamide (23) the nucleic acid is mixed with the support in a soluble state and this is followed by a treatment in which the support is rendered insoluble, thereby physically trapping the polynucleotide strands within the support matrix. For example, the DNA-cellulose acetate complex is formed by dissolving the two materials in an anhydrous solvent and then adding water to co-precipitate the two polymers, while immobilization in polyacrylamide gels is effected by the polymerization of acrylamide in the presence of the polynucleotide. Cellulose-DNA is prepared by simply drying a slurry of DNA and cellulose (24). Although these methods of immobilization have the advantage of simplicity in preparation the complexes formed are subject to dissociation and are useful only for the immobilization of nucleic acids of large molecular weight where the rate of diffusion from the support matrix is relatively slow compared with the time taken to complete a particular chromatographic experiment.

APPLICATIONS

There are three main research areas in which immobilized polynucleotides have been exploited:

1. Fractionation of polynucleotides and nucleic acids through base-paired complex formation.

2. Study of the mechanism of enzymes involved in the synthesis or degradation of nucleic acids.

3. Isolation of enzymes involved in the synthesis or degradation of nucleic acids by affinity chromatography.

Fractionation of Polynucleotides and Nucleic Acids

The base-pairing properties of cellulose containing terminally-linked thymidine oligonucleotides was first demonstrated with a mixture of deoxyadenosine oligonucleotides of chain lengths, 3-7 (1,2). The mixture was fractionated on a column of the oligo(dT)-cellulose using an elution procedure involving a stepwise temperature gradient. The temperature at which each oligonucleotide could be eluted corresponded roughly to the dissociation temperature of the complex formed in solution between the oligonucleotide and thymidine dodecanucleotide. Oligoribonucleotide fragments obtained from viral RNA could also be fractionated on the basis of their content of contiguous adenosine sequences by chromatography on oligo(dT)-cellulose using temperature gradient elution (25).

The initial observation that oligo(dT)-cellulose could be used to specifically isolate A-rich polynucleotides from the total mammalian cellular RNA (26) has led to the widespread use of the material for the isolation and study of messenger RNA species that contain covalently-linked poly(A) segments. Examples of the application of the method include the isolation of poly(A)-containing nuclear RNA of HeLa cells (27), the purification of rabbit globin mRNA (28), the study of the *in vivo* and *in vitro* synthesis of poly(A)-rich RNA by rat brain (29), the purification of the mRNA coding for a mouse immunoglobin L-chain (30), the purification of the 14s mRNA coding for the A_2 chain of α-crystallin of calf lens (31), and the assay of the poly(A) content of yeast mRNA (32).

It should be pointed out that, in the preparation of oligonucleotide-celluloses for the isolation of nucleic acids by the specific base-pairing mechanism, the commercial source of the cellulose used is of some importance. Oligo(dT)-cellulose synthesized from a cellulose preparation that differed from the type originally specified (2) has been shown to be no more efficient in the binding of poly(A) than the unsubstituted cellulose itself (33). A more recent study (34) of the binding of polynucleotides to various celluloses has shown that some cellulose preparations are capable of binding considerable amounts of poly(A) and poly(I) in the absence of derivatization with oligo(dT) or oligo(dC) and it was suggested that this non-specific binding probably results from a relatively high lignin content in these preparations.

Celluloses containing polynucleotides that have been terminally linked by the water-soluble carbodiimide method have also been investigated for their ability to selectively bind polynucleotides containing complementary sequences. A comprehensive study of the preparation and the binding properties of immobilized oligodeoxyribonucleotides of defined chain length and sequence has been carried out and the results indicate that such materials should be useful in the isolation of mRNA species containing particular oligonucleotide sequences (35,36). In another application, the RNA prepared from the DNA of the simian virus, SV40, has been attached to cellulose by the carbodiimide procedure and the product was shown to be capable of specifically absorbing that portion of fragmented SV40 DNA that contained sequences complementary to the

immobilized RNA (37). The technique has some potential as a
general method for gene isolation since the specificity of the
absorption could be maintained even in the presence of a large
excess of bacterial DNA.

Covalent immobilization via the periodate oxidation of the
3'-terminals of polynucleotides has also been used for the isolation
of complementary polynucleotides. Periodate oxidized E. coli 16S
rRNA bound to an agarose derivative containing hydrazide groups
was found to be effective in purifying complementary DNA chains
from a mixture of fragments prepared from sheared E. coli DNA (38).
A recent novel application involves the use of the complex formed
between a periodate-oxidized tRNA species and a polyacrylamide gel
containing hydrazide groups to specifically bind another tRNA
species possessing a complementary anticodon sequence (39). The
technique constitutes a new approach to the study of complementary
anticodons as well as a method for the purification of certain
tRNA species.

The use of polynucleotides immobilized by multiple covalent
linkages for the isolation of nucleic acids containing complementary sequences was first reported by Bautz and Hall (12).
Bacteriophage T4 DNA was attached to acetylated phosphocellulose
and the resulting material was used to chromatographically separate
T4-specific RNA from E. coli RNA. More recent applications of this
separation technique have made use of supports containing polynucleotides that have been attached by the cyanogen bromide or UV-
irradiation methods. Poly(U) attached to cyanogen bromide-activated
agarose has been employed in the isolation of mRNA from KB-cells
(40), and in the study of the poly(A) segments in HeLa mRNA (41,43).

Myeloma cell mRNA containing poly(A) sequences has been
isolated by chromatography on polyvinyl beads containing UV-
irradiated poly(U) (44), and the same absorbent has been used to
effect a partial purification of tRNA^{ser} from crude E. coli tRNA
(45). In addition, poly(U)-cellulose prepared by the UV-irradiation
technique has found use in a number of studies concerned with the
poly(A) sequences in viral and eukaryotic RNA species (21,46,47),
while ribosomal RNA-cellulose prepared by the same method has
served as an absorbent in the partial purification of ribosomal
RNA genes from B. subtilis (20). Merriam et al. (48) have prepared, by the UV-irradiation method, a cellulose complex containing
DNA from the bacteriophage, ØX174, and have used the material for
the analysis of complementary strands in DNA preparations obtained
from cells infected with the phage.

The immobilization of single-stranded DNA by physical entrapment in gels of cellulose acetate or agar produces materials that
are capable of forming specific complexes with RNA molecules containing complementary sequences, and this property forms the basis

of a general method for the isolation of RNA species that possess
sequences that are complementary to DNA from a particular source.
The immobilization of DNA in agar has been subsequently applied to
a number of studies on nucleic acids and the techniques of the
preparation and uses of these materials have been reviewed (49).

Mechanism of Enzyme Action

The methods used for the binding of polynucleotides by
terminal covalent attachment to a support are such that the
orientation of the bound molecules is known, and this property
allows the use of these immobilized polymers in the study of the
action of those enzymes that are associated with the synthesis or
degradation of nucleic acids. Oligo(dT)-cellulose can be used as
a substrate primer for terminal deoxynucleotidyl transferase to
extend the immobilized oligo(dT) chain at its 3'-terminus with a
covalently-linked poly(dC) chain (50). This product together with
soluble poly(dC) is capable of forming a bihelical structure with
a common template molecule, poly(dI), and the resulting complex
may be used to assay and study the enzyme, polynucleotide ligase.
Further studies along these lines have shown that oligo(dT)-
cellulose serves also as a primer and template for E. coli DNA
polymerase and as a template for RNA polymerase (51). For example,
the synthesis of poly(dT)-poly(dA) by DNA polymerase in the presence
of a complex formed between oligo(dT)-cellulose and oligo(dA) yields
a product in which the poly(dT) chain is covalently bound to the
cellulose and the poly(dA) chain is hydrogen-bonded to the poly(dT)
chain.

Polynucleotides with terminal phosphate groups have been
linked to soluble polysaccharides such as Ficoll and dextran using
the water-soluble carbodiimide technique, and the resulting macromolecules have been shown to possess some novel properties in their
use in the study of the action of the enzymes, deoxyribonuclease,
polynucleotide kinase, and DNA polymerase (52). The water-soluble
carbodiimide technique can also be applied to the attachment of
fragmented DNA to an insoluble cross-linked dextran and the resulting complex serves as a substrate for pancreatic deoxyribonuclease
and as a template for RNA polymerase (53). Polynucleotides that
have been linked at multiple points to cyanogen bromide-activated
agarose also act as useful substrates in certain biochemical applications. RNA-agarose prepared in this way forms the basis of an
assay procedure for an endonuclease isolated from sea urchin
embryos (16). This nuclease degrades RNA to large polynucleotides
and is somewhat difficult to assay by the classical methods.
Poly(I)-agarose and the base-paired complex that it forms with
poly(C) have been prepared as potential reagents for the study of
the mechanism of induction of host resistance to viral infection
(15).

Affinity Chromatography

The method of purification of proteins that exploits their capacity to specifically bind to immobilized polynucleotides has found extensive application in the study of those enzymes that bind to DNA. Initial applications of the affinity chromatographic method in nucleic acid research have employed DNA physically immobilized in agarose (54) or cellulose (24), and DNA immobilized on cellulose by UV-irradiation (18). These chromatographic materials have been used for the purification of endonuclease I and exonucleases I and II (54), DNA polymerase (18,24), RNA polymerase (24), and the gene 32-protein of bacteriophage T4 (24). The methods for the preparation and use of DNA-cellulose together with other applications of the chromatographic method have been reviewed by Alberts and Herrick (55). DNA immobilized in polyacrylamide gel has also been used for the purification of DNA polymerase (56). Many of the enzymes that are associated with the synthesis of nucleic acids possess the capacity to bind to single-stranded DNA and the use of single stranded DNA-agarose for the preparative purification of E. coli DNA polymerases I and II, RNA polymerase, exonuclease III, and bacteriophage T4 polynucleotide kinase has recently been described (57).

The covalent attachment of DNA to agarose by the cyanogen bromide procedure also produces materials that are suitable for affinity chromatography. DNA polymerase (14) and deoxyribonucleases from pancreas and spleen (58) have been purified by this method. In two other applications, the complex formed by the UV-irradiation immobilization of bacteriophage f2 RNA on cellulose has served in the chromatographic purification of the poly(G) polymerase obtained from f2-infected E. coli cells (19), and oligo(dT)-cellulose, prepared by the carbodiimide procedure, has provided a purification method for the RNA-dependent DNA polymerase from RNA tumor viruses (59).

Immobilized tRNA has been employed in a number of studies on the isolation of specific aminoacyl-tRNA synthetases. The chromatographic absorbent used in two of these studies was prepared by the periodate oxidation of a particular tRNA species and the subsequent covalent binding of the product to a support containing hydrazide groups. Affinity chromatographic purifications have been effected with $tRNA^{val}$ or $tRNA^{lys}$ bound to a polyacrylhydrazide-agar mixture (60), and with $tRNA^{phe}$ attached to hydrazinyl-agarose (8). In the latter case, the chromatographic method permitted the isolation of yeast phenylalanyl-tRNA synthetase in a completely pure state. Aminoacyl-tRNA may be immobilized by the covalent binding of the amino group of the aminoacyl moiety to an activated support. E. coli isoleucyl-tRNA can be bound to bromoacetamidobutyl-agarose and the resulting complex serves as a chromatographic system for the purification of the corresponding tRNA synthetase (61).

CONCLUSION

It is apparent from the foregoing discussion that, for a particular study, the choice of the method of immobilization depends, to a large extent, on the nature of the subsequent experimental application of the immobilized polymer. As with the use of other immobilized substances some consideration should be given to the stability of the linkage to the support and the steric availability of the attached polynucleotide for specific interactions with molecules in solution. In certain applications involving the use of covalently immobilized polynucleotides the methods that produce terminal attachment would seem to offer some advantages over those employing multi-point attachment. In studies concerned with the assay and the mechanism of action of nucleic acid-associated enzymes the methods employing single point attachment at one of the polynucleotide terminals are desirable since they yield products in which the polarity of the attached polymer is known and in which there is likely to be the least steric interference from the support matrix. In the isolation, by the base-pairing mechanism, of nucleic acids possessing homopolymer sequences, supports containing complementary homopolymers immobilized by single or multi-point attachment seem equally effective. In the future however, more sophisticated separations of nucleic acids based on the base-pairing of more complex complementary sequences will be attempted and, in these cases, the use of synthetic oligonucleotides of defined sequence, attached to a support at their terminals, may be preferred. In this regard, it is of interest that a study of the binding capacity of synthetic oligonucleotides bound to cellulose at their 5'-terminals by the water-soluble carbodiimide method indicates that the entire oligonucleotide, in each case, is apparently available for base-paired complex formation with its complementary nucleotide sequence (36).

Future applications of supports containing covalently-linked polynucleotides are likely to add more emphasis to the importance of the nature of the covalent linkage itself. In the case of multi-point immobilized polymers some applications may require a consideration of the number of attachment points per polynucleotide chain that are formed during the immobilization procedure. For example, RNA-agarose prepared by the cyanogen bromide procedure has been used to study an endonuclease that degrades RNA to large oligonucleotides, and the capacity of the enzyme to release RNA fragments from the support was shown to be markedly dependent on the concentration of the cyanogen bromide used to prepare the RNA-agarose (16). For affinity chromatographic applications physical entrapment is a satisfactory method for the immobilization of those nucleic acids that, by virtue of their size, do not undergo rapid diffusion from the support matrix.

ACKNOWLEDGEMENT

This work was supported by grants GM11518 and GM19395 from the National Institutes of Health.

REFERENCES

1. Gilham, P. T., J. Amer. Chem. Soc. 84, 1311 (1962).
2. Gilham, P. T., J. Amer. Chem. Soc. 86, 4982 (1964).
3. Gilham, P. T., Biochemistry 7, 2809 (1968).
4. Jeffers, J. S. and Gilham, P. T., unpublished experiments.
5. Gilham, P. T., Methods Enzymol. 21, 191 (1971).
6. Brown, D. M. and Read, A. P., J. Chem. Soc., 5072 (1965).
7. Knorre, D. G., Myzina, S. D., and Sandakhchiev, L. S., Izv. Sibirsk. Otd. Akad. Nauk SSSR, Ser. Khim. Nauk, 135 (1964).
8. Remy, P., Birmelé, C., and Ebel, J. P., FEBS (Fed. Eur. Biochem. Soc.) Lett. 27, 134 (1972).
9. Weith, H. L., Wiebers, J. L., and Gilham, P. T., Biochemistry 9, 4396 (1970).
10. Rosenberg, M., Wiebers, J. L., and Gilham, P. T., Biochemistry 11, 3623 (1972).
11. Rosenberg, M. and Gilham, P. T., Biochim. Biophys. Acta 246, 337 (1971).
12. Bautz, E. K. F. and Hall, B. D., Proc. Natl. Acad. Sci. U.S. 48, 400 (1962).
13. Adler, A. J. and Rich, A., J. Amer. Chem. Soc. 84, 3977 (1962).
14. Poonian, M. S., Schlabach, A. J., and Weissbach, A., Biochemistry 10, 424 (1971).
15. Wagner, A. F., Bugianesi, R. L., and Shen, T. Y., Biochem. Biophys. Res. Commun. 45, 184 (1971).
16. Berridge, M. V. and Aronson, A. I., Anal. Biochem. 53, 603 (1973).
17. Britten, R. J., Science 142, 963 (1963).
18. Litman, R. M., J. Biol. Chem. 243, 6222 (1968).
19. Fedoroff, N. V. and Zinder, N. D., Proc. Natl. Acad. Sci. U.S. 68, 1838 (1971).
20. Smith, I., Smith, H., and Pifko, S., Anal. Biochem. 48, 27 (1972).
21. Sheldon, R., Jurale, C., and Kates, J., Proc. Natl. Acad. Sci. U.S. 69, 417 (1972).
22. Bolton, E. T. and McCarthy, B. J., Proc. Natl. Acad. Sci. U.S. 48, 1390 (1962).
23. Wada, A. and Kishizaki, A., Biochim. Biophys. Acta 166, 29, (1968).
24. Alberts, B. M., Amodio, F. J., Jenkins, M., Gutmann, E. D., and Ferris, F. L., Cold Spring Harbor Symp. Quant. Biol. 33, 289 (1968).

25. Gilham, P. T. and Robinson, W. E., J. Amer. Chem. Soc. 86, 4985 (1964).
26. Edmonds, M. and Caramela, M. G., J. Biol. Chem. 244, 1314 (1969).
27. Edmonds, M., Vaughan, M. H., and Nakazato, H., Proc. Natl. Acad. Sci. U.S. 68, 1336 (1971).
28. Aviv, H. and Leder, P., Proc. Natl. Acad. Sci. U.S. 69, 1408 (1972).
29. DeLarco, J. and Guroff, G., Biochem. Biophys. Res. Commun. 49, 1233 (1972).
30. Schechter, I., Proc. Natl. Acad. Sci. U.S. 70, 2256 (1973).
31. Berns, A. J. M., Bloemendal, H., Kaufman, S., and Verma, I. M., Biochem. Biophys. Res. Commun. 52, 1013 (1973).
32. McLaughlin, C. S., Warner, J. R., Edmonds, M., Nakazato, H., and Vaughan, M. H., J. Biol. Chem. 248, 1466 (1973).
33. Kitos, P. A., Saxon, G., and Amos, H., Biochem. Biophys. Res. Commun. 47, 1426 (1972).
34. DeLarco, J. and Guroff, G., Biochem. Biophys. Res. Commun. 50, 486 (1973).
35. Astell, C. and Smith, M., J. Biol. Chem. 246, 1944 (1971).
36. Astell, C. R. and Smith, M., Biochemistry 11, 4114 (1972).
37. Shih, T. Y. and Martin, M. A., Proc. Natl. Acad. Sci. U.S. 70, 1697 (1973).
38. Robberson, D. L. and Davidson, N., Biochemistry 11, 533 (1972).
39. Grosjean, H., Takada, C., and Petre, J., Biochem. Biophys. Res. Commun. 53, 882 (1973).
40. Lindberg, U. and Persson, T., Eur. J. Biochem. 31, 246 (1972).
41. Adesnik, M., Salditt, M., Thomas, W., and Darnell, J. E., J. Mol. Biol. 71, 21 (1972).
42. Jelinek, W., Adesnik, M., Salditt, M., Sheiness, D., Wall, R., Molloy, G., Philipson, L., and Darnell, J. E., J. Mol. Biol. 75, 515 (1973).
43. Molloy, G. R. and Darnell, J. E., Biochemistry 12, 2324 (1973).
44. Stevens, R. H. and Williamson, A. R. Nature (London) 239, 143 (1972).
45. Hung, P. P., Science 149, 639 (1965).
46. Kates, J., Cold Spring Harbor Symp. Quant. Biol. 35, 743 (1970).
47. Philipson, L., Wall, R., Glickman, G., and Darnell, J. E., Proc. Natl. Acad. Sci. U.S. 68, 2806 (1971).
48. Merriam, E. V., Dumas, L. B., and Sinsheimer, R. L., Anal. Biochem. 36, 389 (1970).
49. Bendich, A. J. and Bolton, E. T., Methods Enzymol. 12 B, 635 (1968).
50. Cozzarelli, N. R., Melechen, N. E., Jovin, T. M., and Kornberg, A., Biochem. Biophys. Res. Commun. 28, 578 (1967).

51. Jovin, T. M. and Kornberg, A., J. Biol. Chem. 243, 250 (1968).
52. Scheffler, I. E. and Richardson, C. C., J. Biol. Chem. 247, 5736 (1972).
53. Rickwood, D., Biochim. Biophys. Acta 269, 47 (1972).
54. Naber, J. E., Schepman, A. M. J., and Rorsch, A., Biochim. Biophys. Acta 114, 326 (1966).
55. Alberts, B. M. and Herrick, G., Methods Enzymol. 21, 198 (1971).
56. Cavalieri, L. F. and Carroll, E., Proc. Natl. Acad. Sci. U.S. 67, 807 (1970).
57. Schaller, H., Nusslein, C., Bonhoeffer, F. J., Kurz, C., and Nietzschmann, I., Eur. J. Biochem. 26, 474 (1972).
58. Schabort, J. C., J. Chromatogr. 73, 253 (1972).
59. Gerwin, B. I. and Milstein, J. B., Proc. Natl. Acad. Sci. U.S. 69, 2599 (1972).
60. Nelidova, O. D. and Kiselev, L. L, Molekul. Biol. 2, 60 (1968).
61. Bartkowiak, S. and Pawelkiewicz, J., Biochim. Biophys. Acta 272, 137 (1972).

IMMOBILIZED COFACTORS AND MULTI-STEP ENZYME-SYSTEMS

Klaus Mosbach

Biochemical Division, Chemical Center University of Lund

P. O. Box 740, S-220 07 Lund, 7, Sweden

IMMOBILIZED COFACTORS

Immobilized cofactors or coenzymes are of interest for two major reasons. They show potential as active immobilized coenzymes when used in enzymic processes such as steroid transformation (1); they find use as ligands in biospecific affinity chromatography. The various cofactor analogues synthesized in our laboratory and immobilized to Sepharose include: ε-aminohexanoyl-NAD$^+$ (2), NAD$^+$-N^6[N-(6-aminohexyl)-acetamide], and N^6-carboxy-methyl-NAD$^+$ (3), N^6-(6-aminohexyl)-AMP (4-6) and 8-(6-aminohexyl)-aminocyclic AMP (7).

The first two NAD$^+$-analogues showed some coenzymic activity immobilized to Sepharose of which the N^6-substituted NAD$^+$-analogue appears to be the more active derivative. The various nucleotides were applied as "general ligands" in affinity chromatography. In particular the N^6-substituted AMP analogue was tested and applied successfully in the separation of a number of NAD$^+$-dependent dehydrogenases. The various elution media used included NAD$^+$/NADH (4,6), compounds permitting ternary complex formation (4,6,8) including adducts (9), cofactor (8) and salt-gradients (10). The isozymes of lactate dehydrogenase were separated using an NADH-gradient (11). The cyclic AMP-analogue was applied in the purification of protamine kinase (7).

IMMOBILIZED MULTI-STEP ENZYME-SYSTEMS.

The following multi-step enzyme-systems with the participating enzymes working in sequence have been immobilized on matrices to simulate organization of enzymes likely to be present in situ. Hexokinaseglucose-6-phosphate dehydrogenase (12), β-galactosidase-hexokinase-glucose-6-phosphate dehydrogenase (13) and malate dehydrogenase-citrate synthase-lactate dehydrogenase (14). It was shown that binding of enzymes to the same matrix leads to a "kinetic advantage" over the corresponding soluble system.

Another system studied, amyloglycosidase-glucose oxidase, showed a shift of the pH-optimum of the coupled system compared to the soluble one (15).

The possible role of protons as a regulatory medium was studied with the enzymes hexokinase, glucose oxidase and trypsin immobilized within the same polymer beads (16).

The latter studies led to the development of enzyme-pH-electrodes (17). Other analytical applications resulting from the studies above include the use of immobilized enzymes entrapped in spherical beads of polyacrylamide with defined size (18) in microcalorimetry (19).

REFERENCES

1. Mosbach, K. and Larsson, P. O. (1970) Biotechn. & Bioeng. Vol. XII, 1927.
2. Larsson, P. O. and Mosbach, K. (1971) Biotechn. & Bioeng. Vol. XIII, No. 3, 393-398.
3. Lindberg, M., Larsson, P. O. and Mosbach, K., Eur. J. Biochem. in press.
4. Mosbach, K., Guilford, H., Larsson, P. O., Ohlsson R. and Scott, M. (1972) Biochem. J. 125, No. 2, 20-21.
5. Guilford, H., Larsson, P. O. and Mosbach, K. (1972) Chemica Scripta, 2, 165-170.
6. Mosbach, K., Guilford, H., Ohlsson, R. and Scott, M. (1972) Biochem. J. 127, 625-631.
7. Jergil, B., Guilford, H. and Mosbach, K., Biochem. J. in press.
8. Ohlsson, R., Brodelius, P. and Mosbach, K. (1972) FEBS Letters, 25, No. 2, 234-238
9. Dixon, J. E., Everse, J., Lee, C. Y., Taylor, S. S., Kaplan, N. O. and Mosbach, K., to be published.
10. Lowe, C. R., Mosbach, K. and Dean, P. (1972), Biochem. & Biophys. Res. Comm., 48, No. 4, 1004-1010.
11. Brodelius, P. and Mosbach, K. (1973) FEBS Letters, 35, No. 2, 223-226.
12. Mosbach, K., and Mattiasson, B., (1970) Acta Chem. Scand. 24, 2093-2100.
13. Mattiasson, B., and Mosbach, K., (1971) Biochim. et Biophys.

Acta, 235, 253-257.
14. Srere, P. A., Mattiasson, B. and Mosbach, K., (1973) Proc. Nat. Acad. Sci., 70, No. 9 2534-2538.
15. Gestrelius, S., Mattiasson, B. and Mosbach, K. (1972) Biochim. et Biophys. Acta, 276, 339-343.
16. Gestrelius, S., Mattiasson and Mosbach, K. (1973) Eur. J. Biochem. 36, 89-96.
17. Nilsson, H., Akerlund, A. C. and Mosbach, K. (1973) Biochim. et Biophys. Acta, 320, 529-534.
18. Nilsson, H., Mosbach, R. and Mosbach, K. (1972) Biochim. et Biophys. Acta, 268, 253-256.
19. Johansson, A., Lundberg, J., Mattiasson, B. and Mosbach, K. (1973) Biochim. et Biophys. Acta 304, 217-221.

PREPARATION, CHARACTERIZATION AND APPLICATIONS OF ENZYMES IMMOBILIZED ON INORGANIC SUPPORTS

H. H. Weetall

Corning Glass Works

Corning, New York

The objective of this report is to indicate to the reader some of the parameters which must be investigated when developing an immobilized enzyme system for industrial applications. This report will discuss these parameters utilizing data collected in our laboratories.

The studies described were conducted utilizing inorganic supports of either the porous glass or porous ceramic type as carriers.[1-5] Inorganic supports can be reacted with silane coupling agents as shown in Figure 1 to produce an alkylamine derivative. Silanes will react not only with glass as shown, but also with metal oxides including ZrO_2, TiO_2, Al_2O_3, and NiO.

$$-O-\underset{\underset{O}{|}}{\overset{\overset{O}{|}}{Si}}-OH \quad + \quad NH_2(CH_2)_3 Si(OCH_2CH_3)_3$$

Glass　　　　　　　　　　Silane

$$-O-\underset{\underset{O}{|}}{\overset{\overset{O}{|}}{Si}}-O-\underset{\underset{O}{|}}{\overset{\overset{O}{|}}{Si}}-(CH_2)_3 NH_2 \quad + \quad CH_3CH_2OH$$

Fig. 1. Schematic representation of the coupling reaction between a glass surface and a silane. The silane in this example is γ-aminopropyltriethoxysilane. The product is an alkylamine-glass derivative.

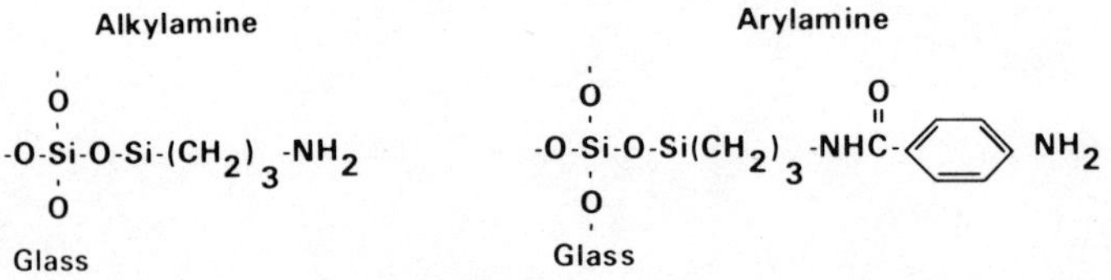

Alkylamine

$-O-Si-O-Si-(CH_2)_3-NH_2$ (with O above and below Si)
Glass

Arylamine

$-O-Si-O-Si(CH_2)_3-NHC(=O)-C_6H_4-NH_2$ (with O above and below Si)
Glass

Fig. 2. Examples of two inorganic silane derivatives most commonly used in our laboratories for the covalent attachment of enzymes.

Enzymes were covalently attached to the inorganic carriers by using the organic functional group of the silanized carrier and an organic group of the enzymes. The two most common groups for covalent attachment used in our laboratories are the aryl- and alkylamines (Figure 2).

Several methods are available for the attachment of the organic functional group on the carrier's surface to an enzyme. Figure 3 shows a method successfully used for the covalent attachment of proteins at acid pH. In this case, the alkylamine is first converted to a carboxylic acid by treatment with succinic anhydride followed by activation with carbodiimide. We have successfully coupled papain by this method,[6] and pepsin to porous glass by a slight modification of the technique.[7]

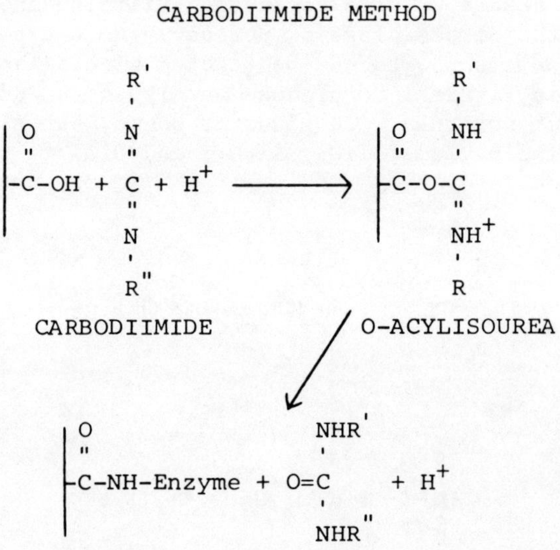

CARBODIIMIDE METHOD

$$-C(=O)-OH + \underset{\underset{R''}{|}}{\overset{\overset{R'}{|}}{N=C=N}} + H^+ \longrightarrow -C(=O)-O-C(=NH)(NH^+R)$$

CARBODIIMIDE   O-ACYLISOUREA

$$\downarrow$$

$$-C(=O)-NH-Enzyme + O=C(NHR')(NHR'') + H^+$$

Fig. 3. The method used for attachment of enzyme at acid pH to an inorganic carboxylated derivative. The carboxyl derivative was prepared by reaction of the alkylamine derivative with succinic anhydride.

AZO METHOD

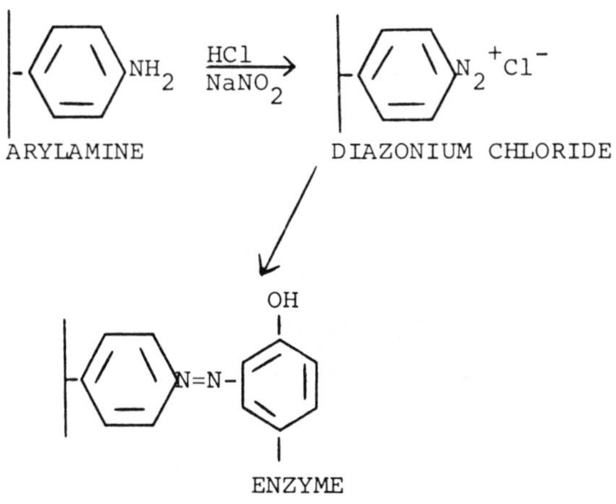

Fig. 4. The method used for covalent attachment of an arylamine derivative to an enzyme. The arlyamine is first diazotized followed by reaction with the enzyme at pH 8-9.

An extremely useful coupling method involves use of the arylamine derivative. This derivative can be characterized and reacted at slightly alkaline pH with enzymes or other proteins (Figure 4). Azo linkage has been most successful and is the method with which we have had the greatest experience.

GLUTARALDEHYDE METHOD

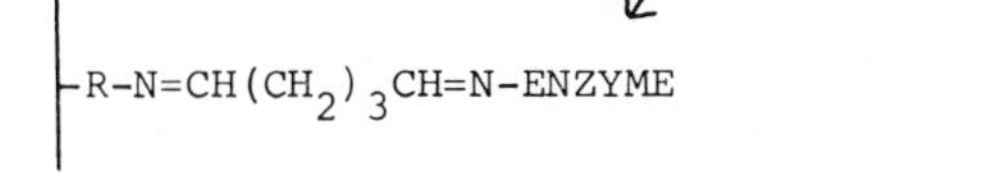

Fig. 5. This is a simplified schematic representation of the reaction between an alkylamine derivative and an enzyme using the bifunctional reagent glutaraldehyde. To prevent cross-linking the excess glutaraldehyde is removed before reaction with the enzyme. The double bonds may be reduced with sodium borohydride.

The covalent attachment of protein to alkylamino derivative can be accomplished via glutaraldehyde (Fig. 5). This is also a simple and relatively gentle method of coupling. To prevent cross-linking we wash out the excess glutaraldehyde after the initial reaction of the bifunctional aldehyde with the carrier. This is followed by attachment of the enzyme to the remaining free aldehyde group. Experience in our laboratory has shown that reduction of the remaining double bonds does not generally increase enzyme half-life. Therefore, for most studies using glutaraldehyde as the coupling agent, the product is not reduced.

One of the first parameters investigated when characterizing an immobilized enzyme is the pH profile. This is necessary in order to determine the usefulness of the enzyme after immobilization since pH shifts on immobilization are the rule rather than the exception. An example of an observed shift in pH profile is given in Figure 6.

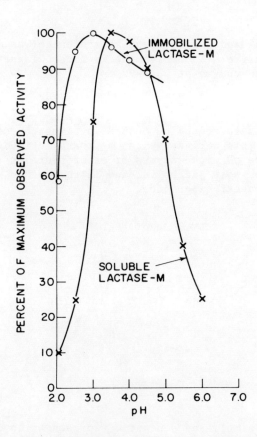

Fig. 6. pH profiles of immobilized fungal lactase compared with soluble lactase. Assays were carried out in lactose substrate. (From <u>Biotechnology and Bioengineering</u>, IN PRESS)

The lactase purchased from Miles Laboratories was covalently coupled to ZrO₂-coated porous glass. The pH profile on attachment to the carrier shifted from an optimum of pH 4.0 to pH 3.0. Since our major interest at this time was the development of a system for the treatment of acid whey, the shift was acceptable. Figure 6 shows that at pH 4.5-5.0 the immobilized enzyme still gives 80% to 90% of the maximum observable activity. We have examined other lactases as well. Figure 7 gives the results observed with a fungal lactase obtained from Wallerstein Laboratories. The enzyme was covalently coupled to ZrO₂-coated porous glass and to a TiO₂ porous ceramic. The Figure shows that the glass derivative is the shifted farthest to the acid side. The soluble enzyme has a pH optimum between pH 5-6. These data indicate that it may be possible to intentionally shift the pH optimum of an immobilized enzyme at will by proper

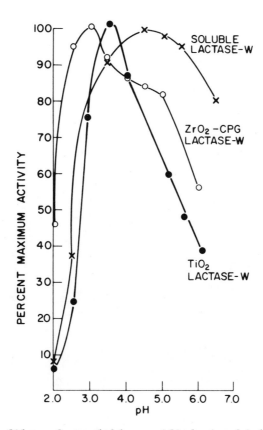

Fig. 7. pH profiles of a soluble purified microbial lactase compared with the same enzyme covalently coupled to ZrO₂ coated porous glass and a TiO₂ porous ceramic assays were carried out in lactose substrate (From Biotechnology and Bioengineering, IN PRESS)

choice of carrier. Figure 8 gives the pH profile of a third lactase. In this case, the pH optimum shifted 3.5 pH units such that a completely useless enzyme at pH 4.0 is now showing 80% to 90% of maximum activity at pH 4.0.

Another parameter of importance is thermal stability. Figure 9 gives the Arrhenius plot of the Wallerstein lactose immobilized on ZrO_2-porous glass and a TiO_2 porous ceramic. The observed decreasing velocity observed on the porous glass sample at increased temperature is not denaturation, but the effect of pore diffusion.

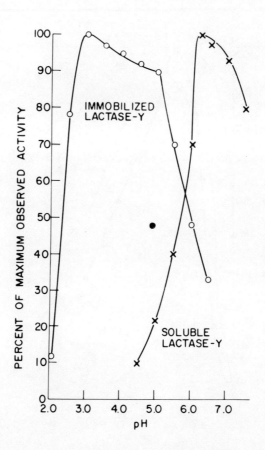

Fig. 8. pH profile of a soluble yeast lactase compared with the same enzyme covalently coupled to ZrO_2 coated porous glass. Note the extreme shift in pH optimum (From Biotechnology and Bioengineering, IN PRESS)

At the higher temperatures the enzyme turnover rate exceeds the diffusion rate of substrate and product, thus velocity decreases due to substrate limitation and/or product inhibition.

The kinetics of an immobilized enzyme are also very important parameters. For our studies, the apparent kinetic values are sufficient for all necessary calculations. Figures 10-11 are typical examples of Lineweaver-Burk type plots for soluble and immobilized lactases used for the calculation of $K_{m(app)}$ and $K_{i(app)}$ for galactose. Figures 12-13 summarize the data from Figures 10-11.

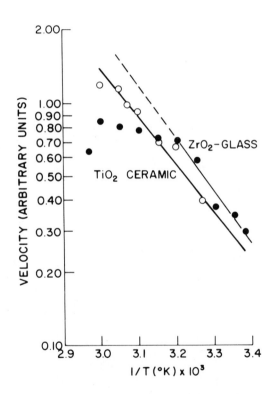

Fig. 9. Arrenhius Plot of immobilized purified fungal lactase on ZrO_2 coated porous glass and on TiO_2 porous ceramic particles. The effect of pore diffusion is observed on the ZrO_2 particles as shown by the decrease in velocity with increasing temperature. The lack of apparent diffusion control with the TiO_2 particles is also due to mass transfer problems. (From Biotechnology & Bioengineering, IN PRESS)

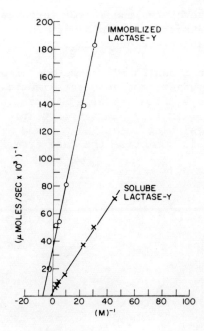

Fig. 10. Lineweaver – Burk plot of a soluble and immobilized yeast lactase. The enzymes were assayed in lactose substrate. The carrier was ZrO_2 coated porous glass.

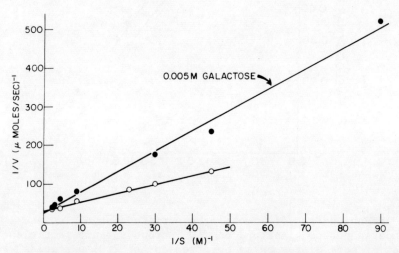

Fig. 11. Inhibition plot of purified fungal lactase covalently coupled to ZrO_2 coated porous glass with the addition of 0.005M galactose compared to the same derivative assayed in the absence of the inhibitor.

The results indicate that although the K_m and K_i values of the soluble enzymes are different, immobilization equalizes these values. The activation energy appears well within the expected range. The low value may be due to pore diffusion.

ENZYME PREPARATION	$K_{m(app)}$ (M)	$K_{i(app)}$ (MM)	Ea (Kcal/mole)
LACTASE-Y (SOLUBLE)	0.112	0.666	10.5
LACTASE-Y (IMMOB.)	0.0714	3.27	11.3
LACTASE-M (SOLUBLE)	0.04	1.25	n.d.
LACTASE-M (IMMOB.)	0.05	3.22	6.54

Fig. 12. Kinetic parameters of soluble and immobilized lactase. All derivatives were immobilized on ZrO_2-coated porous glass. (From Biotechnology and Bioengineering, IN PRESS)

ENZYME PREPARATION	K_m (M)	K_i (MM)	E_a (Kcal/g mole)
LACTASE-M (SOLUBLE)	0.04	1.3	-
LACTASE-M (IMMOBILIZED ON ZrO_2-CPG)[1]	0.05	3.2	6.5
LACTASE-W (SOLUBLE)	0.05	3.9	-
LACTASE-W (IMMOBILIZED)[1,2]	0.07	3.8	7.8

[1] APPARENT K_m and K_i

[2] WITHIN 95% CONFIDENCE LIMITS THERE WAS NO SIGNIFICANT DIFFERENCES BETWEEN THE K_m AND K_i OF THE ZrO_2-CPG AND THE POROUS TiO_2.

Fig. 13. Kinetic parameters of immobilized lactase compared with the soluble enzyme. All derivatives were prepared on ZrO_2-coated porous glass. (From Biotechnology and Bioengineering, IN PRESS)

Generally, the literature on immobilized enzyme technology indicates that $K_m(app)$ of an immobilized enzyme is larger than that of the soluble enzyme. The data presented in Figure 14 indicate cases where no change in K_m occurs, instances with increased K_m values, and decreased K_m values. At this time we have no way of predicting the effect immobilization will have on K_m.

With the aid of the kinetic data we can now develop some models and make some predictions as to requirements for an immobilized enzyme system. Figure 15 gives the computer plots of lactase units

ENZYME	SUBSTRATE	K_m (millimolar)	
		SOLUBLE	IMMOBILIZED
INVERTASE	SUCROSE	0.448	0.448
ARYLSULFATASE	p-NITROPHENYL-SULFATE	1.85	1.57
GLUCOAMYLASE	STARCH	1.22	0.30
ALKALINE PHOSPHATASE	p-NITROPEHNYL-PHOSPHATE	0.10	2.90
UREASE	UREA	10.0	7.60
GLUCOSE OXIDASE	GLUCOSE	7.70	6.80
L-AMINO ACID OXIDASE	L-LEUCINE	1.00	4.00

THE ENZYMES WERE IMMOBILIZED ON POROUS 96% SILICA GLASS PARTICLES

Fig. 14. Comparison of K_m values of some soluble and immobilized enzymes prepared on ZrO_2 coated and uncoated porous glass derivatives.

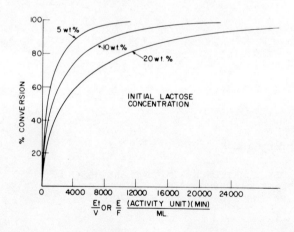

Fig. 15. Computer representation of enzyme quantity required to achieve a given percent hydrolysis of lactose as determined by use of experimentally derived kinetic parameters and the rate equation. (From <u>Biotechnology and Bioengineering</u>, IN PRESS)

required vs percent lactose hydrolyzed at three different substrate concentrations. By utilizing this type of curve we can determine the quantity of enzyme required to build a plant capable of producing any quantity of product per unit time. The curve also tells us that as percent lactose hydrolyzed increases, the quantity of enzyme required increases drastically. Thus, to produce 90% hydrolyzed lactose requires over ten times the quantity of enzyme as 50% hydrolysis assuming quantity of lactase treated per unit time is constant.

Once having the computer model we must experimentally test the models validity. Figure 16 gives the results of both batch and

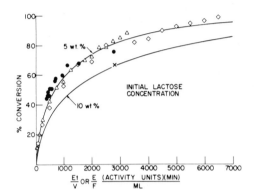

Fig. 16. Experimental verification of computor plot shown in Figure 15 by reaction in fixed bed reactors ◇, ○ and batch reactor △. (From Biotechnology and Bioengineering, IN PRESS)

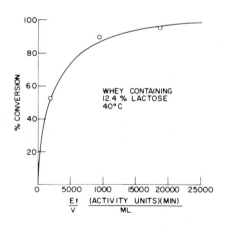

Fig. 17. Enzyme required for 12.4% lactose substrate as determined from rate equation compared with data experimentally determined using concentrated whey ultrafiltrate. (From Biotechnology and Bioengineering, IN PRESS)

column reactor studies compared against the computer plots. The experimental data do indeed verify the model. Figure 17 is a similar plot for a whey ultra filtrate concentrate which indicates the model holds for purified lactose and lactose in whey.

Once we have validated the model we can now employ the curve for an additional purpose, that of determining enzyme half-life. A detailed description of the method is published elsewhere.[6,9] The curve in effect tells us how much enzyme is in a column operating at constant flow rate. This is accomplished by determining enzyme activity directly from percent hydrolysis observed. The enzyme activity can then be treated by regression analysis and a half-life determined.

Figure 18 gives some half-life data for immobilized lactases versus operating temperature. At higher temperatures we observe decreased half-lives. The half-life of the lactase-M appears by far the highest and looks quite promising. However, the preparation contains only one-tenth the activity of the other enzyme preparations. Figure 19 gives the results of studies carried out on whole sweet whey, acid whey, and acid whey ultrafiltrate. The next figure (Figure 20) gives examples of typical observed half-lives of other enzymes immobilized on inorganic carriers. The extremely high half-life observed for glucoamylase appears to be a real value, but extremely unusual and quite unique when compared with other immobilized enzymes.

ENZYME	TEMP. (°C)	DAYS OF OPERATION	MEAN $t_{\frac{1}{2}}$ (DAYS)	UCL* (DAYS)	LCL* (DAYS)
LACTASE-M	30	55	198	465	126
"	40	35	38.3	48.2	31.8
LACTASE-W (ZrO_2-CPG)	30	36	43.9	63.0	33.6
"	40	24	34.7	47.0	27.4
"	50	4	2.7	4.9	1.7
" (NaB$_4$ TREATED)	40	23	35.8	55.9	26.4
LACTASE-W (POROUS TiO_2)					
35/45 MESH	40	33	39.8	57.7	30.1
1 mm particles	40	37	77.9	93.6	69.3
1 mm particles	40	37	66.6	92.4	53.3
1 mm particles+	40	45	58.2	96.3	41.3

*THE CORRELATION OF THESE DATA WITH EXPONENTIAL DECAY IN ALL CASES EXCEEDED 0.85.

+THESE DATA WERE OBTAINED FROM A COLUMN OPERATED AT pH 6.5.

Fig. 18. Half-life studies on immobilized lactase derivatives using lactose substrate (From Biotechnology and Bioengineering, IN PRESS)

ENZYME PREP.	TEMPERATURE (°C)	SUBSTRATE	OPERATION TIME (DAYS)	HALF-LIFE (DAYS)
POROUS TiO_2	40	SWEET WHEY	35	24.9[+]
"	40	WHEY ULTRAFILTRATE	34	24.5[+]
"	40	ACID WHEY	34	28.5[+]
"	40	SWEET WHEY	46	53.7[++]

[+]THESE VALUES WERE DETERMINED BY CALCULATION OF $t_{\frac{1}{2}}$ USING ONLY INITIAL AND FINAL ENZYME ACTIVITIES OF THE DERIVATIVE.

[++]THIS VALUE WAS DETERMINED BY REGRESSION OF THE DAILY RESULTS AS PREVIOUSLY DESCRIBED.[1] THE 95% CONFIDENCE LIMITS ON THIS VALUE WERE UCL-82.5 DAYS, LCL-39.4 DAYS. pH WAS MAINTAINED at pH 5.2.

Fig. 19. Half-life studies on immobilized lactase derivatives using cheese whey as substrate.

ENZYME	TEMP. (°C)	SUBSTRATE	HALF-LIFE (DAYS)
L-AMINO ACID OXIDASE	37	L-LEUCINE	43
ALKALINE PHOSPHATASE	23	p-NITROPHENYL-PHOSPHATE	55
PAPAIN*	45	CASEIN	35
LACTASE (YEAST)*	50	LACTASE	20
GLUCOAMYLASE*	45	STARCH	645

*THE POROUS GLASS USED IN THESE STUDIES WAS ZrO_2 COATED, 40-80 MESH, 550 Å PORE DIAMETER.

Fig. 20. Half-lives of enzymes immobilized on porous glass.

Half-life of enzymes under operational conditions is dependent, to a great extent, on operating temperature. Figure 21, is an excellent example of the effect of temperature on the half-life of immobilized glucoamylase. At 60°C the operational half-life of the derivative is 15 days; however, as the operating temperature is decreased, the half-life increases. Thus, at 40°C the half-life is 900 days. Another point of interest is the decrease in reaction rate vs temperature. Taking the rate of 60°C as 100%, we observe a 75% decrease in reaction rate at 40°C; however, the greatly increased half-life still means that the immobilized enzyme is more economically operated at 40°C than at 60°C.

One can plot half-lives vs reciprocal of temperature in an Arrhenius type plot (Figure 22). These data permit one to extrapolate half-lives to higher or lower temperatures. Thus by plotting the Wallerstein enzyme data and extrapolating to 10°C we estimate an apparent half-life in excess of 2 years.

Additional parameters of importance include optimal surface area and pore diameters of the enzyme carrier. Figure 23 gives an excellent example of these effects on the activity of an immobilized glucoamylase. It is immediately apparent that a pore diameter of approximately 550 Å is optimal for this enzyme using an enzyme thinned cornstarch as substrate.

TEMPERATURE (°C)	HALF-LIFE (DAYS)	RELATIVE REACTION RATE (% OF THAT AT 60°C)
60	13	100
50	100	70
45	645	30
40	900	25

Fig. 21. Half-life of glucoamylase immobilized on ZrO_2 coated porous glass compared with operating temperature and relative velocity.

ENZYMES IMMOBILIZED ON INORGANIC SUPPORTS

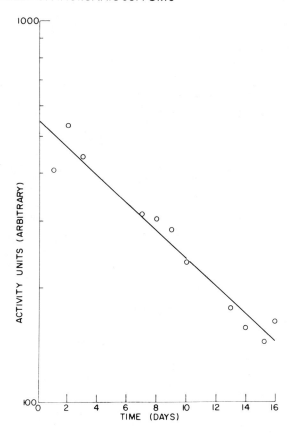

Fig. 22. Regression analysis of column half-life data showing experimental decay of enzyme activity vs time.

AVG. PORE DIAMETER, Å	SURFACE AREA, m^2/g	ACTIVITY (UNITS)
2300	8	502
1380	12	723
875	29	1950
550	60 - 80	3000
248	80 - 120	50

Fig. 23. The effect of pore volume on the activity of immobilized glucoamylase using enzyme thinned cornstarch as substrate.

The choice of a reactor type can also make a great difference in the economic evaluation of an immobilized enzyme. Figure 24 gives an example of enzyme requirements vs percent hydrolysis for immobilized glucoamylase using a plug flow reactor and a continuous stirred tank reactor (CSTR). At 95% hydrolysis the CSTR requires a much greater quantity of enzyme to produce an equivalent amount of product as the plug flow reactor.

The results of these types of studies can be summarized as shown in Figure 25. Thus, a series of three columns at 40°C, each one 6" x 8' in size, operating on 30% solids, enzyme thinned cornstarch, could produce 10,000,000 lbs. of dextrose/yr. The cost of operating such a plant would be 10-50% less than present day costs depending upon plant size.

At Corning we have not only explored the industrial potential of immobilized enzymes, but we have also studied these materials for analytical applications.

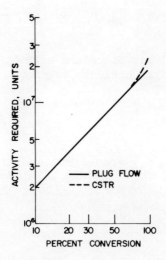

Fig. 24. Comparison of reactor enzyme requirements for an immobilized glucoamylase using a plug flow and a continuous stirred tank reactor (CSTR).

One system we have developed uses immobilized alkaline phosphatase for the detection and quantitation of inorganic phosphate.[10] The enzyme hydrolyses p-nitrophenylphosphate to p-nitrophenol plus inorganic phosphate. The addition of inorganic phosphate to the reaction mixture shifts the equilibrium decreasing the quantity of p-nitrophenol produced. Since p-nitrophenol is green while the substrate is colorless, we can monitor the change colorimetrically. The system used consisted of a programmed 2-part rotary valve which allowed the sequential addition of up to 20 different solutions and a vessel containing substrate and a flow switch which permitted fluid to pass through the column or around the column for base line determinations. The reaction was monitored by a flow-through spectrophotometer (Figure 26). If the system is operated in a high ionic strength buffer any nonspecific anionic effect is eliminated and the system reacts specifically to added inorganic phosphate. Figure 27 shows a typical standard curve obtained using this system.

SUBSTRATE: ENZYME THINNED CORNSTARCH OF 30% SOLIDS

CARRIER: ZrO_2 COATED POROUS GLASS 18-20 MESH

pH: 4.5

TEMPERATURE: 60°C (IF A 40°C SYSTEM WERE USED, THEN 7.5 CU.FT. OF IMMOBILIZED ENZYME WOULD BE REQUIRED)

SPECIFIC ACTIVITY: 3000 UNITS/G (ONE UNIT PRODUCES 13.8 mg DEXTROSE/hr. AT 60°C)

K_m: 3×10^{-4} M

REACTOR TYPE: PLUG FLOW COLUMN

REACTOR SIZE: APPROXIMATELY 2.5 CU.FT. OR 6" dia. x 8' height

SUBSTRATE RESIDENCE TIME: 5.9 MINUTES

PRESSURE DROP ACROSS COLUMN: 40 psi

Fig. 25. Summary of parameters for a glucoamylase pilot plant for the production of 10,000,000 lbs. (30% D.S.) of 95% dextrose syrup per year.

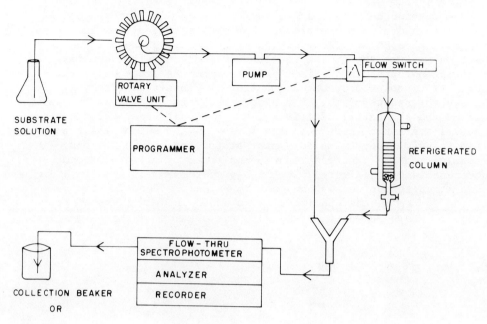

Fig. 26. Schematic representation of apparatus used for the detection and quantitation of inorganic phosphate using immobilized alkaline phosphatase. The system consists of a 20 port rotary valve capable of feeding several unknown and known samples of inorganic phosphate into the reactor. The valve is programmed permitting the unknown is mixed with substrate as it flows to the column. For background determination a solanoid valve feeds the substrate in a bypass around the column. Product is monitored in a flow-through colorimeter.

The reciprocal of percent hydrolyses is proportional to added phosphate concentration. Results obtained with this standard curve on some solutions containing added phosphate are shown in Figure 28.

A third type of application for immobilized enzymes at Corning is therapeutic. Dr. L. S. Hersh of our laboratory has carried out a series of studies utilizing L-asparaginase convalently coupled to polymethacrylate plates.[11] These plates have been used in an extracorporeal shunt for the hydrolysis of L-asparagine to aspartic acid in human volunteers. The unit is attached at one end to the arterial flow and at the other to the venous flow. The blood pressure forces blood through the shunt and back into the body.

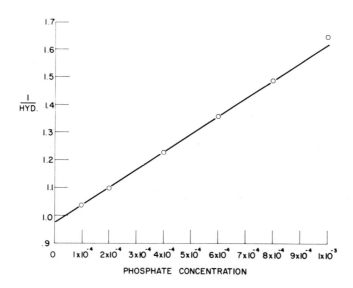

Fig. 27. Standard curve for quantitation of inorganic phosphate using immobilized alkaline phosphative with p-nitrophenylphosphate as substrate.

ADDED COMPONENTS TO TEST SOLUTION	ACTUAL PHOSPHATE CONCENTRATION	EXPERIMENTAL PHOSPHATE CONCENTRATION
None	3.0×10^{-4} M	2.6×10^{-4} M
	9.0×10^{-4} M	9.0×10^{-4} M
.001 M NaNO$_3$ .01 M Na$_2$SO$_4$	3.0×10^{-4} M	2.9×10^{-4} M
	9.0×10^{-4} M	8.9×10^{-4} M
River Water	3.0×10^{-4} M	3.0×10^{-4} M
	9.0×10^{-4} M	9.2×10^{-4} M

Fig. 28. Determination of phosphate concentration in samples of solution having known quantities of inorganic phosphate.

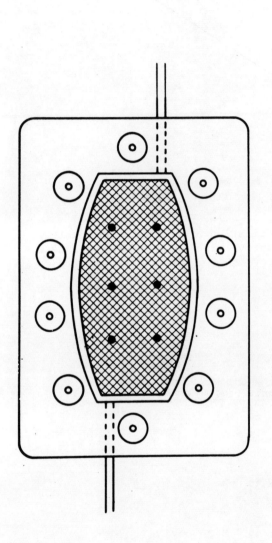 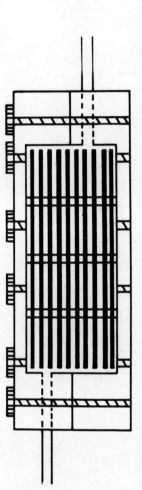

Fig. 29. Schematic of extracorporeal shunt developed by L. S. Hersh. Plates are of poly-methacrylate and are coupled with enzyme through a silane coupling agent.

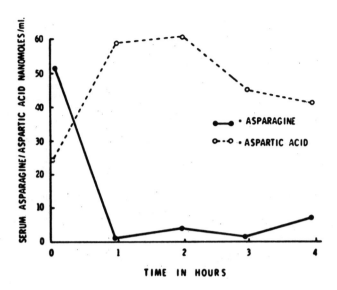

Fig. 30. Results observed on attachment of shunt to human volunteer.

The graph illustrates the rapid fall in serum asparagine levels following commencement of the extracorporeal perfusion, and comensurate rise in serum aspartic acid is seen during the first 2 hours. After this, however, due to the asparagine depletion, serum aspartic acid levels fall again. This pattern was consistently seen in all the experiments.

Figure 29 gives a schematic of the unit. Results with this shunt are shown in Figure 30. We note that the L-asparagine decreases to a minimal level while the aspartic acid rises. These results were obtained in a human volunteer who was exposed to the shunt for four hours. The experiments were carried out at Roswell Park Memorial Hospital on kidney dialysis patients.[12] The results showed that it is indeed possible to treat patients with immobilized enzymes in the form of an extracorporeal shunt.

This report has described some of the parameters investigated in our attempts to develop an immobilized enzyme system for industrial application. The parameters mentioned are only a small portion of the many parameters one studies before any immobilized enzyme system can even be economically evaluated. However, it should give the reader some indication of the quantity of work and the many details which must be examined in commercially immobilized enzymes. This report has also shown that there are other potential uses for immobilized enzymes including analytical and therapeutic applications.

Bibliography

1. Weetall, H. H. and Hersh, L. S., Biochem. Biophys. Acta., 185, 464 (1969).
2. Weetall, H. H., Science, 166, 615 (1969).
3. Weetall, H. H. and Baum, G., Biotech. Bioeng., 12, 399 (1970).
4. Weetall, H. H. and Havewala, N. B., Biotech. Bioeng., Symp. No. 3, 241 (1972).
5. Weetall, H. H., Res. and Develop., 22, 18 (1971).
6. Mason, R. D. and Weetall, H. H., Bioeng. and Biotech., 15 455 (1973).
7. Line, W. F., Kevong, A. and Weetall, H. H., Biochem. Biophys. Acta., 242, 194 (1971).
8. Weetall, H. H., Havewala, N. B., Pitcher, W., Detar, C. C., Vann, W. P. and Yaverbaum, S.
9. Weetall, H. H. and Jacobson, M. A., Fermentation Tech. Today edit. G. Terui, Japan 361-365, 1972.
10. Hersh, L. S., J. Polymer Sci. (In press).
11. Hersh, L. S., Personal Communication.

LACTASE IMMOBILIZED ON STAINLESS STEEL AND OTHER

DENSE METAL AND METAL OXIDE SUPPORTS

M. Charles, R. W. Coughlin, B. R. Allen,
E. K. Paruchuri and F. X. Hasselberger

Department of Chemical Engineering
Lehigh University, Bethlehem, Pa. 18015

INTRODUCTION

The past decade has seen the development of a large number of techniques for immobilizing enzymes which have been reviewed and discussed at length elsewhere(1,2,3). For the present purposes we consider those techniques involving the attachment of enzymes to particulate supports by means other than physical entrapment. While a number of methods have been described, (1,2,3), covalent binding seems to have gained the most favor. Catalysts prepared by such techniques have typically been used in packed bed reactors to process what might be referred to as "reagent grade" substrate streams.

While much has been learned about immobilized enzymes by the use of these binding techniques and packed bed reactors, further developments will probably be necessary before they can be used for practical commercial applications. Typical covalent binding techniques are intricate, time consuming, and labor intensive. Furthermore, they require the use of relatively expensive reagents (many of which are toxic) and solid supports. Those solid supports which are not expensive tend to be either mechanically inferior (e.g., soft gels, filter paper) or subject to microbial attack (e.g., natural polymers). The packed bed reactor requires high pressure drops for reasonable throughputs and is easily plugged. It is, therefore, expensive to operate and difficult to maintain in continuous operation. These problems increase in magnitude as the size of catalyst particles decreases. High rates of reaction and large degrees of conversion are, however, favored by the use of small particles. As will be discussed below, fluidized beds are capable of conversions at least as great as those obtainable from comparable packed beds and are not subject to the mechanical limitations associated with packed beds.

Our purpose in this paper is to describe the operation and advantages of fluidized beds in exploiting the characteristics of immobilized enzyme catalysts and to discuss the preparation and characteristics of such catalysts which we have developed specifically for use in fluidized beds. Results of packed and fluidized bed experiments using lactase bound to stainless steel, nickel oxide and alumina supports are presented.

FLUIDIZED BEDS; CHARACTERISTICS AND ADVANTAGES

During fluidization a bed of solid particles is transformed into a fluid-like state by passing a fluid upward through the bed at a velocity which is sufficiently large to overcome the net gravitational force acting on the particles but sufficiently small to prevent the particles from being transported out of the containing vessel(4,5). When there is no fluid flow, the bed merely rests on the base of the column. As fluid flows through the column with increasing velocity, eventually a point is reached at which the pressure drop multiplied by the cross-sectional area of the bed becomes equal to the bed weight. At this point, the particles separate slightly from each other and begin moving slowly about. The bed takes on the appearance of a gently boiling liquid and exhibits many properties characteristic of liquids. As the flow rate is increased beyond that required to first induce fluidization, the separation of the particles becomes more pronounced (the bed expands) and the motion of the bed becomes more vigorous. Continued increase of the flow rate will eventually sweep all of the particles from the column.

Although solid particles can be fluidized by either gases or liquids, gas-fluidized beds are generally the more common in industrial applications. In a gas-fluidized bed, much of the gas tends to pass through the bed in the form of bubbles. The rapid rise of these bubbles promotes violent and rapid mixing of the bed particles and as a result the bed has excellent heat and mass transfer characteristics. In addition, once the particles are in a fluidized state they can be pneumatically transported out of the bed with relative ease. Clearly, this is quite desirable when the particles must be removed from the bed for regeneration.

Such advantages of fluidized-bed operation are well known. However, there are many recognized <u>disadvantages</u> of <u>gas</u>-fluidized beds. The violent motion within a gas-fluidized bed can cause a considerable amount of elutriation (particles are literally hurled from the free surface of the bed by gas bubbles breaking the surface) and it also promotes particle attrition which can further increase elutriation. This violent motion can also cause extensive wearing of equipment. In addition, the presence of gas bubbles and the mixing they cause can introduce problems related to chemical conversion (6) since the gas within a bubble, which contacts virtually no solid, will pass through the bed essentially unreacted (bypassing). Yet another disadvantage of gas-solid fluidized beds is caused by backmixing of the solids. Such backmixing can have a deleterious effect on conversion, particularly when it is the solid that is being treated in the reactor. Furthermore, the backmixing of solids can induce backmixing of gas which can lead to decreased conversions when the gas is the material being treated.

The particle motion in a liquid-fluidized bed is very much gentler than that in a gas-fluidized bed. It has been found (7) that, although "bubbles" (of liquid) exist in liquid-fluidized beds, they are very much smaller than those found in gas-fluidized beds. It appears that, for reasons of stability, the liquid bubbles are limited in size by the size of the bed particles. Thus, the smaller the bed particles, the smaller the bubbles and hence the more uniform and smoother the fluidization (<u>i.e</u>., the "quality" of fluidization is improved). This behavior has indeed been observed experimentally (7). The radical difference between gas-fluidized beds and their liquid counterparts is probably best appreciated by observing the free surfaces of both types. In a gas-fluidized bed, the free surface is poorly defined with large aggregates of particles being frequently "blown out" of the main body of the bed by the gas bubbles as they break the surface. In a liquid-fluidized bed, on the other hand, the free surface is quite sharp and well defined and eruptions by bubbles do not occur.

Since the particle motion in liquid-fluidized beds is not violent, attrition and elutriation problems are considerably reduced. This is evidenced by the well-defined demarkation between the bed and the solid-free

liquid above it. Despite the diminished chaos, in a
liquid-fluidized bed, one can readily observe that the
particles are well mixed. It has also been found (8)
that liquid backmixing is not very great in such beds
and that the liquid motion through the bed can be very
close to plug flow in character. In essence, the liquid
flow through a liquid-fluidized bed can be made quite
similar to that through a packed bed and this is a major
advantage for liquid-fluidized beds over CSTR's for
carrying out reactions catalyzed by insolubilized
enzymes.

There are several other factors which are found to
affect the fluidization quality of liquid-fluidized
beds (7). Increasing the difference in density (for a
constant particle size) between the liquid and solids
will tend to diminish the quality of fluidization whereas
increasing the liquid viscosity will tend to enhance it.
Generally, some compromise will be required. For example, higher liquid flow rates are desirable since they
result in better mass transfer and increased production
rates. Increasing the density difference between liquid
and solid will permit higher liquid velocities without
the particle terminal velocity being exceeded. The increase in density difference, however, tends to decrease
the quality of fluidization and hence some compromise is
necessary. It might be well to emphasize at this point
that enzymes may be attached to materials of widely differing densities and hence the compromise alluded to has
real meaning.

In heterogeneous reactions such as the type being
considered in this paper, the first step in the reaction
sequence is the diffusion of the substrate from the
liquid phase to the exterior surface of the solid particle. If the solid is nonporous, the next step is the
chemical reaction. If the solid is porous, the second
step will be the diffusion of substrate molecules through
the pores of the solid. Both of the diffusion steps can
often be rather slow in comparison to the enzyme-substrate reaction itself and hence almost anything that
can be done to decrease the diffusive resistance without
significantly decreasing the driving force for diffusion
will tend to increase the overall (global) rate of reaction. The simplest and most effective way to decrease
both these resistances simultaneously is to decrease the
size of the solid particles. This increases the outer
surface area of the particle which decreases the first
diffusional resistance and decreases the average length
of the pores through which the substrate molecules must

diffuse and thereby decreases the second resistance (the internal surface is effectively brought closer to the outside). Furthermore, the decreased particle size provides a greater reactive surface area per unit volume of reactor.

An important advantage of fluidized beds is that they can operate with very small particles ($\sim$ 100 microns) without plugging. Furthermore, as was previously noted, the use of smaller particles results in fluidization of better quality. The only precaution that need be taken is to ensure that the density of the particles is sufficiently high to permit reasonable liquid flow through the column without exceeding the terminal velocity of the particles. As has been previously remarked, enzymes can be bound to materials having a sufficiently wide range of density that enzyme-bearing solids could be prepared with the proper size and density to suit virtually any set of reasonable fluidization conditions.

Packed beds of small particles in contrast to fluidized beds, are generally not found to be suitable for the treatment of industrial liquid streams. The primary reason for this is that such beds can act as efficient filters (consider the typical sand bed filter used in water purification processes) and there is generally a sufficiently high level of suspended particulate in many industrial streams to threaten rapid plugging of packed beds of small particles. It is interesting to note that it is this problem that stimulated the development of fluidized-bed ion-exchange processes (9,10,11). One such process (11) has been demonstrated to have the ability to handle slurries with solids loadings of at least 10 percent. In general, then, only relatively large particles ($\sim$ 1/4 inch) can be used as catalysts in most practical packed columns for streams of typical "industrial" purity.

In connection with the use of small particles it should also be noted that for the same throughput, the pressure drop across a packed bed is very much larger than that across a fluidized bed. Furthermore, while the pressure drop across a packed bed increases markedly with increasing flow rate, that across a fluidized bed is virtually insensitive to flow rate. The disparity increases quickly as the particle size decreases.

The advantages of the fluidized bed discussed above are not the only ones which suggest its use. Another important advantage of the liquid-fluidized bed is that it provides for simple and continuous withdrawal and addition of solids (e.g., immobilized enzyme catalysts) which may therefore be regenerated on a continuous basis. This permits maintenance of a constant bed activity for an indefinite period of time and hence obviates the necessity of continually adjusting production rates; this presupposes, of course, that economical continuous regeneration processes can be developed. In a fixed packed bed, continuous withdrawal and regeneration is difficult if not impossible. As time goes by, the catalyst activity will slowly decrease and this in turn will require a decrease in production rate. Furthermore, a point will eventually be reached at which the activity will have decreased to an unacceptably low value. At this point, the packed bed must be taken off stream so that the catalyst can be removed for regeneration. If production is to be continuous, a second column will obviously be required. In such multiple column operation, a common example of which is the fixed-bed adsorption process, a large inventory of catalytic solids is required and an appreciable part of this inventory remains idle for relatively long periods of time. This is particularly disadvantageous when the cost of the solid catalyst is high as is the case for most bound-enzyme systems. This inventory problem could be circumvented for fluid-bed processes if regeneration of solid, immobilized enzyme catalysts could be made continuous.

Another possible advantage of fluidized-bed operation is the ease of maintaining a constant and uniform temperature. Maintenance of a uniform temperature in packed beds of any size, on the other hand, can be difficult. Although large heat effects are not associated with enzyme reactions, it is known that the rates of such reactions are highly temperature dependent and show a very well defined maximum as a function of temperature. It is conceivable that the ability to maintain a constant and uniform temperature throughout the reactor could have a significant effect on the overall conversion. Furthermore, if more than one reaction occurs, even minor temperature variations might have considerable effect on selectivity.

Up to this point, we have compared only fixed- and fluidized-bed reactors. To complete the discussion, we must also consider stirred tank (slurry) reactors. With regard to mass transfer, it has been found by Calderbank (12) that the mass transfer coefficient (k_m) for the transfer of substrate to a solid particle in a CSTR or slurry reactor is of the same order as the transfer coefficients for fluidized and packed columns. In addition, very small particles (even smaller than those used in fluidized beds) can be used in a slurry reactor. One might therefore expect that such reactors would always compete quite favorably with fluid-bed reactors. This, however, is not the case since both solid and liquid phases are completely mixed in a CSTR and hence the overall average driving force for mass transfer throughout the reactor will be lower for the CSTR than it will be in the plug-flow pattern that exists in the fluidized bed, assuming identical feed conditions. This decrease in driving force will tend to decrease conversion. It must be remarked, however, that some care must be exercised when comparing these reactor types. For example, for the same liquid rate and the same weight of catalyst (same activity and size) the relative conversions will depend on the height-to-diameter ratio chosen for the fluidized bed. Other factors, such as particle size, must also be considered. Similar care should be taken when comparing slurry reactors to packed-bed reactors.

It should also be remarked that slurry reactors are more mechanically complex than either fixed- or fluid-bed reactors and this tends to make them less desirable. Furthermore, the rather intense motion and high shear stresses at impellers and in the vicinity of baffles could lead to appreciable attrition and deterioration of valuable catalyst in CSTR's. In addition, problems of plugging and low liquid-handling capacities are not uncommon (13).

Up to this point, we have limited our discussion to those cases in which the chemical reaction rate is very much greater than the rate of diffusion. As already stated, liquid backmixing in such cases can markedly decrease the overall driving force and hence limit conversion. This was seen to be one of the disadvantages of the slurry reactor. If the rate of chemical reaction is of the same order or slower than the diffusion rate, then solids backmixing will also decrease the driving force and decrease conversion (6). Clearly, then, the slurry reactor would seem to be a particularly poor choice for this case. Also, since there is considerable solids mixing in the fluidized bed and none in the packed bed, the packed bed would seem to be a better choice here.

On the basis of conversion efficiency and ease of operation, both packed bed and fluidized-bed reactors appear to be superior to CSTR's for utilization of immobilized enzyme catalysts. The choice of fluidized over packed beds is based primarily on mechanical considerations. It seems reasonable to expect that for the same particle size and operating conditions both would give approximately the same conversion so long as solids backmixing is unimportant (this is usually the case for immobilized enzyme catalyst). The fluidized bed, however, requires a lower pressure drop, is free of plugging problems, and exhibits all the other advantages described above.

CATALYST PREPARATION

From the preceeding discussion it should be clear that fluidized bed catalysts should be small, dense particles in order to insure minimum mass transfer resistance, a large surface area per unit volume of reactor and good quality of fluidization. The requirement of high density constrains one to consider inorganic materials ranging from oxides of specific gravity 2.5 to metals of specific gravity 19. This, however, is not disadvantageous since such materials possess greater mechanical integrity than the organic supports which have been used in the past and are essentially immune to microbial attack. Most of these inorganic supports are also considerably less expensive than the porous glass support which has been used by many investigators. It might also be noted that the porous glass currently available exhibits very poor fluidization characteristics. The solids we have chosen for study so far include stainless steel (non-porous), nickel-nickel oxide sinter (slightly porous), and an industrial catalyst composed primarily of alumina (porous).

We have attached enzymes to these supports by a number of techniques, but have emphasized two which are effective, simple, and inexpensive. Both of these involve modifications and improvements of techniques which have been described elsewhere (14,15,16,17). In the first, the solid support is "activated" by coating it with titanium oxide [LNOT method]. The "activated" solid is then contacted with an enzyme solution. In the second, the enzyme is first adsorbed onto the support and then crosslinked with glutaraldehyde. To our knowledge, this is the first time that such methods have been used to attach enzymes to the supports noted above. We also believe that this is the first time enzymes have been successfully attached to stainless steel and alumina.

The first technique has been used to bind lactase to nickel oxide, stainless steel and alumina and to bind trypsin and catalase to NiO and to alumina and asparaginase to stainless steel (18). The activities of several of these are given in Table 1.

The second method has been used to prepare, among others, a lactase-stainless steel (LSSG) and lactase-alumina (LAG) catalysts.

TABLE 1

IMMOBILIZED ENZYME CATALYSTS BASED ON TITANIUM OXIDE-TREATED NICKEL OXIDE AND ALUMINA

ENZYME	SUPPORT	ACTIVITY
Lactase	Alumina	7100+
"	"	7020
"	"	910
"	"	2120
"	NiO	910
"	"	796
"	"	1010
"	"	1270
Trypsin	Alumina	1475‡
"	NiO	3.12
Catalase	Alumina	3060.0‡
"	NiO	28.0

+Activity of lactase in Wallerstein units. One Wallerstein unit is equivalent to the amount of enzyme required to convert 3.16×10^{-8} moles of lactase per minute at 37°C, at a pH of 4.5 and at an initial substrate concentration of 20%.

‡One unit of trypsin activity will hydrolyze 1µ mole of TAME per minute at 25°C and at a pH of 8.1 in the presence of 0.01 M calcium ion.

‡One unit of catalase activity decomposes 1µ-mole of hydrogen peroxide per minute at 25°C, at a pH of 7.0, and at an initial substrate concentration of 0.059 H_2O_2.

CATALYST CHARACTERISTICS

The effects of pH and temperature on the activities of lactase-stainless steel (LSSG) and lactase-alumina (LAG) catalysts prepared by the "glutaraldehyde method" are illustrated in Figures 1 and 2. All activities are reported in Wallerstein units. (See Table 1 for the definition of a Wallerstein unit.) These profiles are quite similar to those reported for other immobilized enzymes. It should be noted that the activity of the alumina-based catalyst is very much greater than that of the catalyst based on stainless steel. We feel that this is primarily a result of the fact that the alumina has a greater surface area and is less dense than the

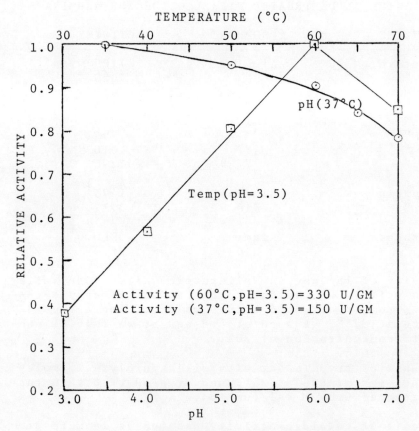

FIGURE 1. TEMPERATURE AND pH PROFILES FOR AN LSSG-TYPE CATALYST.

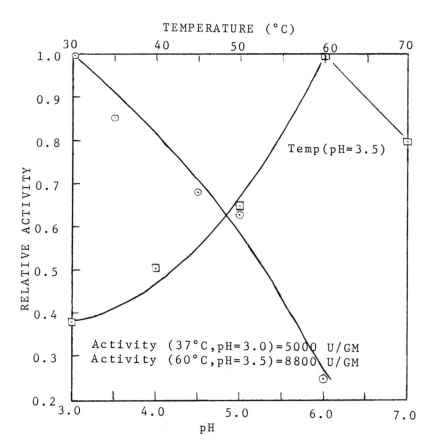

FIGURE 2. TEMPERATURE AND pH PROFILES FOR AN LAG-TYPE CATALYST.

stainless steel. There is also a possibility that alumina adsorbs lactase more strongly. This is currently under investigation.

The effect of "on-stream" time on the activity of an LSSG-type catalyst is shown in Figure 3. The data reported were obtained by using a continuously operated packed bed reactor. The reactor was 10 inches long and had an inside diameter of 0.375 inches. The 1% lactose substrate solution flowed through it continuously at a rate of 0.67 CC/min. It can be seen from these results that these inexpensive, easy-to-prepare catalysts possess good long-term, on-stream stability. Most of our

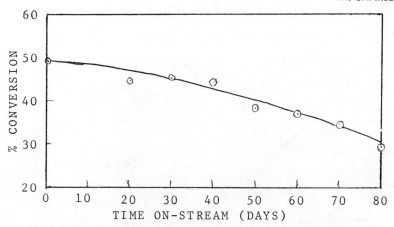

FIGURE 3. ON-STREAM LONG-TERM ACTIVITY OF AN LSSG-TYPE CATALYST.

catalysts demonstrate long-term activities which are as good or better than that of the LSSG catalyst. One LNOT catalyst has maintained its initial activity for 1400 hours on-stream.

Finally, we should note that although we have not as yet determined the binding efficiencies for the methods described we have reason to believe that they are at least comparable to those of other techniques described in the literature.

REACTOR EXPERIMENTS

A lactase-stainless steel catalyst prepared by the glutaraldehyde method (LSSG) and a lactase-NiO catalyst prepared by the $TiCl_4$ (LNOT) method have been used in packed and fluidized bed reactors in order to obtain comparative data for the two reactor types operated under essentially identical conditions. For each catalyst, the same reactor and catalyst charge were used to obtain comparative data for packed and fluidized modes of operation. Packed bed operation was insured by operating in the downflow mode while fluidized bed operation was achieved by operating in the upflow mode. In the case of fluidized bed operation, continuous recycle was used. Schematic diagrams of the systems used are given in Figure 4.

Comparative data for an LNOT catalyst is presented in Figure 5 in which conversion of a lactose substrate is plotted against the number of activity units charged to the bed divided by the substrate volumetric flow rate. The number of activity units, E, is defined as:

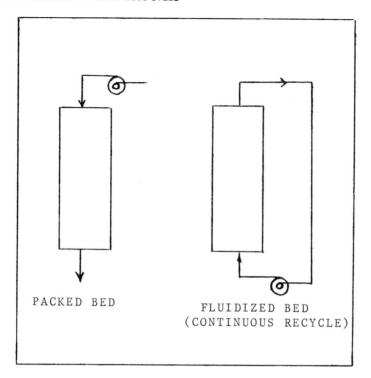

FIGURE 4. REACTOR CONFIGURATIONS USED IN EXPERIMENTAL WORK

$$E = W \cdot e_o \quad (1)$$

where

W = catalyst mass (grams)
e_o = initial specific activity of catalyst (units/gram)

It should be noted that e_o is determined soon after the catalyst is prepared and is generally higher than the specific activity of the catalyst at the time of use. Furthermore, it is determined at a standard temperature of 37°C. It is seen therefore, that E is only an indicator of the actual "activity content" of the reactor and not equal to it. The actual activity could be calculated by applying suitable corrections for temperature and aging. Such corrections would clearly be necessary for design purposes. For purposes of catalyst evaluation and reactor comparisons, however, E is an adequate measure of activity.

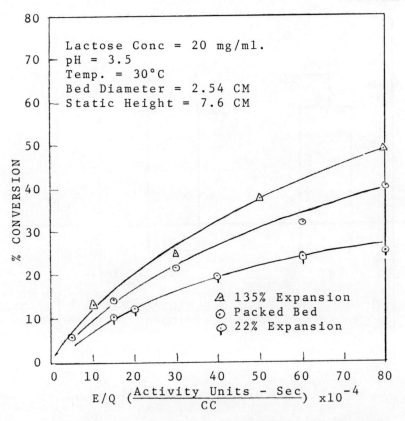

FIGURE 5. CONVERSION VS. E/Q FOR PACKED AND FLUIDIZED BEDS UTILIZING AN LNOT-TYPE CATALYST.

The results presented in Figure 5 demonstrate two important points. First, the fluidized bed conversions are as great or greater than those of the packed bed when the fluidized bed is operated under conditions of good quality fluidization. Second, the fluidized bed conversions are markedly affected by the bed expansion. It is believed that, at low bed expansion, the quality of fluidization is poor which results in a great deal of channeling. Such channeling has been visually observed. At high bed expansions fluidization quality increases and channeling is greatly diminished. From our results it would appear that at bed expansions of the order of 100% the channeling in the fluid bed is no greater than that in the packed bed when an efficient liquid distributor is used in the former. Significant reductions in channeling could probably be achieved by altering the column design (e.g., a better distributor).

It should be noted that the comparative data presented in Figure 5 were all obtained during the same 8 hour period and hence the effect of aging on catalyst activity was negligible.

Fluidized bed results for an LSSG-type catalyst are presented in Figure 6. The results and interpretations are similar to those given for the data of Figure 5. It should be noted that the conversions achieved with this catalyst were considerably lower than those obtained with the preceeding catalyst. There are several important reasons for this. The stainless steel catalyst used was considerably older than the NiO catalyst nor was its activity lifetime as prolonged. Secondly, the stainless steel catalyst was used at 25°C

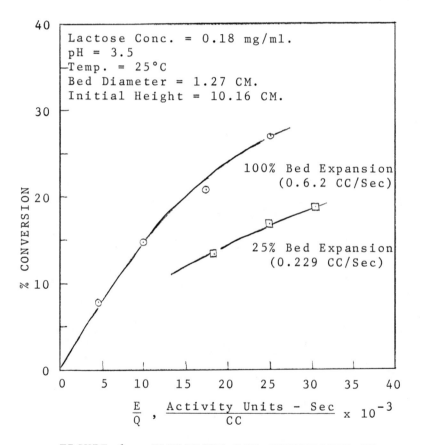

FIGURE 6. FLUIDIZED-BED CONVERSION VS. E/Q FOR AN LSSG-TYPE CATALYST AT TWO DIFFERENT BED EXPANSIONS.

whereas the NiO catalyst was used at 30°C. When these factors are considered it is recognized that the actual activity of the stainless steel catalyst is overestimated by E to a greater extent than is the actual activity of the NiO catalyst.

Lastly, the two catalysts were used in different reactors. The reactor utilizing the stainless steel catalyst is believed to have provided poorer quality fluidization and hence a greater degree of by-passing. For a given value of E/Q, (Q is the volumetric flow rate of substrate solution) a reactor subject to a high degree of by-passing will provide a lower conversion than an identical reactor exhibiting a low degree of by-passing.

Finally, a lactase-alumina catalyst prepared by the glutaraldehyde method (LAG) was evaluated in a fluidized bed. The results are presented in Figure 7. The column

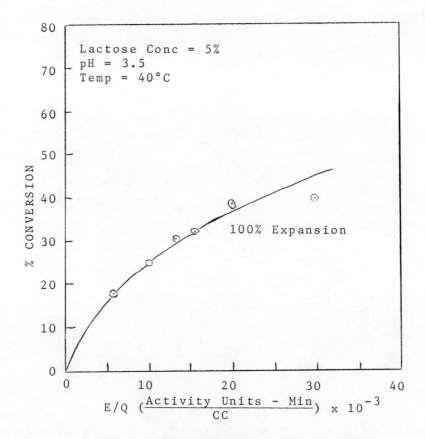

FIGURE 7. FLUIDIZED BED CONVERSION VS. E/Q FOR AN LAG - TYPE CATALYST.

performed well and gave high conversions of a 5% lactose substrate (equivalent to the lactose concentrations of typical dairy waste and product streams). It should be noted that the conversions obtained for a given E/Q were very much less than optimum. The primary reason for this was the use of a new liquid distributor which resulted in rather severe channeling. The distributor, which was designed to provide plug-free operation for particulate containing substrates (e.g., raw whey) is illustrated in Figure 8.

Experiments were also performed to determine the effect of liquid-film mass transfer resistance on conversion. The results indicate that at any value of E/Q the conversion is not affected by the linear flow rate of the substrate solution. From this one may conclude that liquid-film mass transfer resistance is not the controlling factor in the overall rate of reaction.

All of our results (except those in which appreciable channeling was observed) demonstrate this behavior and all are consistent with a model which is

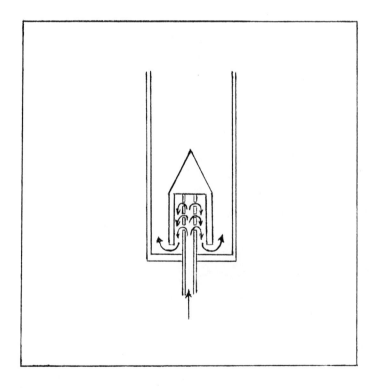

FIGURE 8.   PLUG-FREE LIQUID DISTRIBUTOR.

based on the assumption that the reaction rate and not the mass transfer resistance is limiting. If one substitutes the rate equation for a product inhibited enzyme reaction into the material balance equation for a plug flow reactor, one obtains

$$E/Q = \frac{S_o}{V'_M}\left[1 - \frac{K'_M}{K'_i}\right]X - \frac{K'_M}{V'_M}\left[1 + \frac{S_o}{K'_i}\right]\ln(1-X) \quad (2)$$

where X is the fractional conversion, S_o is the initial substrate concentration and the constants K'_M, V'_M, and K'_I, of the reaction rate equation have their usual meanings. (The primes are used to indicate that these values apply to the bound enzyme.) Dividing this result by X, one obtains

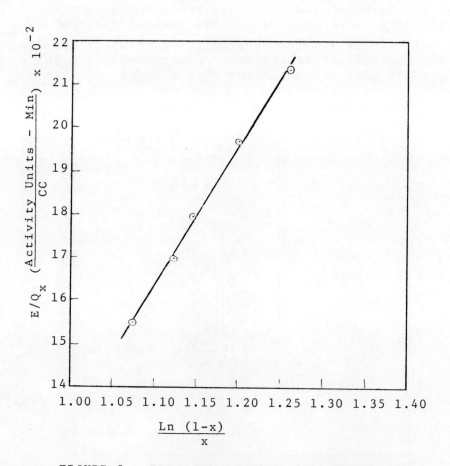

FIGURE 9. PLOT OF EQUATION (3) FOR AN LAG CATALYST USED IN A FLUIDIZED BED OPERATED AT 100% EXPANSION.

LACTASE IMMOBILIZED ON STAINLESS STEEL

$$\frac{E}{QX} = \alpha + \beta \frac{\ln(1-X)}{X} \qquad (3)$$

where

$$\alpha = \frac{S_o}{V'_M}[1 - \frac{K'_M}{K'_i}]$$

$$\beta = \frac{K'_M}{V'_M}[1 + \frac{S_o}{K'_i}]$$

A plot of E/QX vs. $\frac{\ln(1-X)}{X}$ will therefore be a straight line if the model provides an adequate description of the process. Such a plot for a fluidized bed utilizing an LAG-type catalyst is given in Figure 9.

Clearly, one may use the standard techniques of enzyme kinetics to obtain the individual constants K'_M, V'_M, and K'_I but these require the accurate evaluation of reaction rates. Such determinations are time consuming and subject to appreciable error if considerable care is not exercised. While a knowledge of the individual

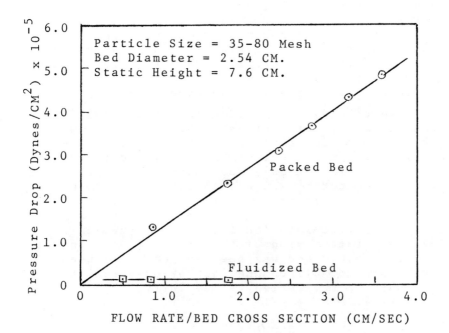

FIGURE 10. PRESSURE DROP VS. FLOW RATE FOR PACKED AND FLUIDIZED BEDS CONTAINING LNOT-TYPE CATALYST.

values of the constants would undoubtedly be useful in extending our understanding of the behavior of immobilized enzyme behavior, they are not necessary for engineering design purposes. So long as equation (3) is applicable, a knowledge of α and β is adequate. As can be seen from the preceeding development these can be readily obtained from integral-reactor type studies which do not require the accurate evaluation of reaction rates.

Finally, experiments were performed to demonstrate, with one of our catalysts, the fact that the pressure drop across a fluidized bed is very much lower than that across the same reactor operated as a packed bed at the same throughput. The results are given in Figure 10.

CONCLUSION

We have presented experimental results which demonstrate that inexpensive, simple-to-prepare immobilized enzyme catalysts can be used in fluidized bed reactors to provide conversions comparable to those obtained with the same catalysts used in packed beds operated under comparable conditions. For some of our catalysts used in fluidized beds conversions as high as those reported by other workers using more exotic catalysts in packed beds were obtained. Arguments presented here as to the mechanical superiority of fluidized beds over packed beds also support the contention that the fluidized bed should be the reactor of choice for utilization of enzymes immobilized on particulate supports.

We have already developed more active catalysts of the type discussed here but have not as yet evaluated them in fluidized bed reactors. We are also presently engaged in the development of better liquid distributors which we feel will not only substantially decrease channeling but also not be subject to plugging. Such distributors will result not only in greater conversions but will also make it possible to process streams containing high particulate loadings (e.g., curd in raw whey). The first version of this type of distributor was described above and has been successfully used (although considerable channeling was observed) to process raw whey at the Lehigh Valley Dairy, Inc., Allentown, Pa. (19).

SUMMARY

Lactase has been bound to very small, dense metal and metal oxide particles such as stainless steel and nickel oxide by simple techniques which permit in situ preparation and regeneration. The catalyst particles have been used in fluidized and packed bed reactors to treat pure lactose solutions and various dairy products including cottage cheese whey.

Side by side comparisons of both reactors have demonstrated that the fluidized bed is the reactor of choice for immobilized enzyme applications of the type studied. In all cases considered fluidized bed conversions were as great as or greater than conversions obtained from packed beds operated under identical conditions of temperature, catalyst mass, residence time, and input concentration. Under conditions of good fluidization good conversions were obtained at low temperatures. Furthermore, fluidized beds exhibited very much lower pressure drops and were not subject to plugging problems so frequently encountered in packed beds.

In this paper we describe methods of catalyst preparation, catalyst characteristics, results of reactor studies. In addition, some comments are made regarding the use of our techniques for the binding and utilization of other enzymes.

ACKNOWLEDGEMENTS

We are grateful for the skillful technical assistance of Mrs. Sue Barbella, Mrs. Jo Gallagher and Mr. William J. Kelly, Jr. Typing was done by Mrs. C. Hildenberger. This work was supported by NSF Grant GI-35997.

REFERENCES

1. Falb, Biotech. and Bioeng. Suppl. 3, 177 (1972).
2. Weetall and Messing, The Chemistry of Biosurfaces, Volume 2, Marcel Dekker, New York, 563 (1972).
3. Chibata and Testuya, Ann. Rept. Tanake Seiyaku Co., Ltd., Tech. Part, Vol.16, 59 (1968).
4. Kunii and Levenspiel, Fluidization Engineering, John Wiley and Sons, Inc., New York, 1969. Chapter 1.
5. Zenz and Othmer, Fluidization and Fluid-Particle Systems, Reinhold Publishing Corp., New York, 1960. Chapter 2.
6. Levenspiel, Chemical Reaction Engineering, John Wiley and Sons, Inc., New York, 1962, pp.295-296.
7. Harrison, Davidson, and de Kock, Trans. Inst. Chem. Engrs., 39, 202 (1961).
8. Bruinzeel, Reman and Van der Laan, Proc. Symp. on The Interaction of Fluids and Particles, London, 1962, pp. 120-126.

9. Grimmett and Brown, Ind. and Eng. Chem., 54, 24 (1962).
10. Selke and Bliss, Chem. Eng. Prog., 47, 529 (1951).
11. Swinton and Weiss, Aust. J. Appl. Sci., 6, 98 (1955).
12. Calderbank and Jones, Trans. Inst. Chem. Engrs., 39, 363 (1961).
13. Satterfield, Mass Transfer in Heterogeneous Catalysis, M.I.T. Press, Cambridge, Mass., 1970, p.107.
14. Barker, Emery and Novais, Proc. Biochem., 6,(10), 11 (1971).
15. Emery, Hough, Novais and Lyons, The Chem. Engr., No. 259, 79 (Feb., 1972).
16. Haines and Walsh, Enzyme Envelopes on Colloidal Particles, Biochem.Biophys.Res.Commun.36, 235-242 (1969).
17. Private Communication.
18. Hasselberger, Charles, Coughlin, Allen and Paruchuri, submitted to BBRC.
19. Coughlin, Charles, Allen, Paruchuri and Hasselberger, paper presented at 66th Annual Meeting AIChE, November 15, 1973, Philadelphia, Pa. (to be published).

THE USE OF MEMBRANE-BOUND ENZYMES IN AN IMMOBILIZED ENZYME REACTOR

Charles C. Worthington

Worthington Biochemical Corporation

Freehold, New Jersey 07728

Our initial interest in immobilized enzymes developed from one particular project: conversion of milk lactose. Consequently the enzyme primarily with which we concerned ourselves was β-galactosidase, and reactor design considerations took into account the requirements of the dairy industry. The result is a basic reactor configuration which lends itself particularly well to applications involving viscous or colloidal substrates and where thorough cleansing must be performed frequently.

Early work with columns was unsatisfactory for, while a small bench top reactor could effect about 80% hydrolysis with good flowrate in respect to column size, bacterial contamination would begin cutting into column efficiency within about a week, even running at 3 - 5°C. No effective method was found for satisfactorily cleaning the column without unpacking the bed. We found similar cleaning problems to apply to other reactors in which substrate turbulence is a function of flow through a restricted passage.

Since turbulence is needed to assure proper substrate-enzyme contact, a lactase membrane was developed which could withstand the shear forces of being stirred in milk. In August, 1970, a simple reactor consisting of a collodion membrane sandwiched between two nylon screens and wrapped around a stirring rod was put into operation. This reactor will be referred to as 8-21.

For 18 months 8-21 was recycled twice per day with one liter aliquots of milk starting each Monday. On Fridays it was washed in cold water, 0.1M $NaHCO_3$, and 70% ethanol and left drying in the cold room for the weekend. The only substrate ever used was fresh skimmed milk. After 18 months milk was recycled only once a day.

Reactor performance is monitored by analyzing timed samples for the glucose formed by the reaction. Since lactose concentration in milk is generally between 4.6 and 5.0 percent, we have assumed completion when the glucose concentration reaches about 2.3%, that is 23 mg/ml. Plotting concentration versus time gives a fairly smooth first order curve until glucose concentration reaches 18 mg/ml. Then the rate begins slowing quite noticeably. Membrane activity is calculated as the initial glucose concentration change per hour times the volume of substrate used. Graph I shows a plot of the average of all points taken for 8-21 during its first three months of operation.

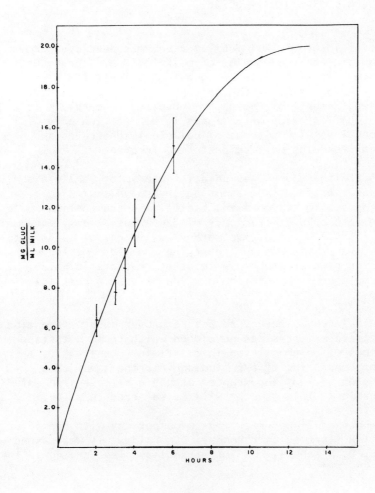

GRAPH I

Graph II is a plot of the average of all data points taken during the twenty-first through twenty-fourth months; there was about a 35% loss in activity. After the two years of continuous operation the activity started dropping rapidly. Graph III is a plot of 8-21's activity each month. By the end of its third year only about one sixth of the original activity remained so use was discontinued.

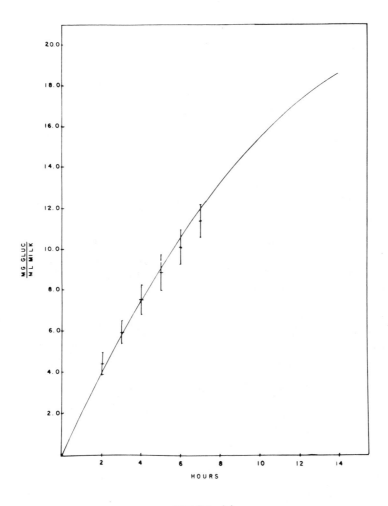

GRAPH II

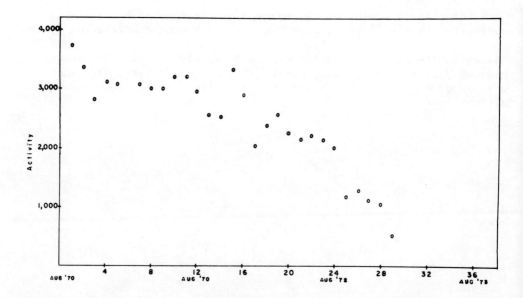

GRAPH III

During this three year period similar devices were assembled. Generally speaking they all performed as well as 8-21; however, larger coiled reactors offered greater difficulty in cleaning. In order to eliminate hidden portions of the membrane but still retain the ability to keep it in motion, we began shaping the membranes as discs. The basic configuration is a circular membrane with the bound enzyme sandwiched between two stainless steel screens. The steel screens serve two purposes: they give the membrane rigidity while the mesh of the screen breaks up the laminar flow around smooth membranes, causing localized turbulence.

Our primary interest at this point was still the milk application, so in designing a reactor we kept in mind the need for easy cleaning, unrestricted flow, and scale-up potential. The result is a column shaped reactor with multiple discs mounted on a central axle; each disc rotates in its own isolated compartment. The substrate enters the compartment at the center axle, flows over the bottom of the rotating disc, across the top, and exits at the center axle at the top of the compartment. The advantages we have found with this reactor design over columns and fluidized beds are summarized below:

Comparison of types:

ENZYMACTOR	MIXED BED BATCH REACTOR	COLUMN
Can handle multiple enzyme system either mixed or sequentially.	Mixed enzymes only.	Sequential or mixed in reactive order.
Analysis is possible for any point of reaction, time or site.	Possible at any time.	Possible only at beginning and end.
Continuous flow or batch.	Batch only.	Continuous flow or batch.
Substrate or co-enzyme can be introduced at any time or point, pH altered, probe inserted at any point.	Limits and imprecision of batch system.	Closed system, no possible changes or modifications can be made.
Substrate and reactant can be introduced or removed at will. Variety of methods possible.	Irrelevant where substrate goes in. Remove bound enzyme by filtration, centrifugation, etc. Possible loss of enzyme without sophisticated separation system.	
Substrate can be viscous. No blocking or plugging of system.	Substrate can be viscous.	Substrate must not be viscous; protein can be a problem.
Membranes do not interfere with each other. Enzymes do not interact or crossreact.	Beads can be crushed physically. Possibility of one enzyme digesting another.	Columns of beads can settle and plug. Possibility of enzymes reacting with each other.
Cleaning-by running fluid through or opening and flushing with running water.	Beads must be separated for cleaning; continuous filtering problem.	Same for mixed bed plus loss of layering in enzymes in column.
Faulty discs can be removed and replaced. Self-contained unit; self-mixing; pump only accessory needed.		

The largest disc reactor we have made thus far uses up to 39 six inch discs and is presently processing five gallons of milk per day allowing one hour for cleaning. The reactor column is made of aluminum coated with a teflon-like polymer. All other parts except the discs are made of 316 stainless steel. Cleaning is performed by releasing four ring clamps, separating the column sides, and hosing with cold tap water. Presently we are considering scaling up to a reactor which can handle twelve inch discs.

When we began using the disc membranes, it became evident that the concept is applicable to a large variety of membrane materials and enzymes. Of particular interest now are applications involving the breakdown of large molecules such as proteins and nucleic acids. We have successfully bound trypsin and chymotrypsin covalently to both nylon and aminoethyl cellulose using glutaraldehyde. Pepsin has been bound to nylon with carbodiimide. Because the reactions occur on the surface and there is no need for the substrate to pass through pores, it has been possible to process viscous solutions of casein and hemoglobin with no difficulty. Most work is done in a batch environment where the disc is simply stirred in a beaker with 100 ml of substrate and the reaction is monitored with the appropriate assay system. For example, with protein substrates, samples are precipitated with 5% trichloroacetic acid and absorbance read at 280nm. When a satisfactory method of binding is indicated, a set of discs is prepared which are loaded into a bench scale reactor and run continuously until there is a significant fall-off in activity. The system is then cleaned thoroughly and restarted. Such procedure can be continued until a fairly good idea of how practical the discs are in terms of quantity of work derived as a function of the cost incurred in making the discs.

At this point we wish to optimize the performance of those enzymes of greatest interest to industry or research. We have expanded our work considerably beyond lactase; however, other applications are only in early stages of development.

We are also evaluating potential modifications in reactor design to better suit industry. Such work includes further simplification of cleaning procedures and reduction of void volume. We feel the basic concept offers considerable potential for moving the immobilized enzyme concept out of the research lab and into industry.

THE OPTIMIZATION OF POROUS MATERIALS FOR IMMOBILIZED

ENZYME (IME) SYSTEMS

        David L. Eaton

        Corning Glass Works

        Corning, New York 14830

Critical parameters involved in the selection of support materials in the immobilization of biologically active materials are discussed. The importance of physical properties such as pore morphology, surface area and particle size relative to the final activity of the enzyme is shown using glucoamylase IME as a model system. The process of selection of specific materials with respect to the environmental conditions of the enzyme and the chemical durability of the support is considered, as are the mechanical properties required in the scale up of an application.

Many different types of materials have been employed as carrier or support materials for the immobilization of enzymes and other biologically active materials. These range from organic polymers and cellulose-type materials to inorganic supports like clay, charcoal, silica gel and controlled-pore glass (CPG). They cover a very broad spectrum of materials, but few have the capability of being selectively optimized for a specific application.

In optimizing a carrier material for use with a specific application, like using the enzyme glucoamylase to convert starch to glucose, many considerations must be made. The first and most obvious is that we are dealing with an inorganic carrier and an organic catalyst composite (Fig. 1). Each part of the composite system imposes restraints on the other. Practically speaking, however, it is the inorganic half which is most often modified or selected because the enzyme is generally very labile in

```
                    R
                    O
                    |
    C.P.G.-O-Si-R-N=N~~~                    → Glucose
                    |
                    O
                    R
```

CARRIER - COUPLING AGENT - ENZYME + SUBSTRATE → PRODUCT

$$\text{STARCH} \xrightarrow{\text{Glucoamylase}} \text{GLUCOSE}$$

APPLICATION

Figure 1: Depiction of immobilized enzyme composite consisting of an inorganic support material, a silane coupling agent and the active enzyme. The specific case shown is for glucoamylase converting polymeric starch molecules to a simple sugar glucose.

nature and cannot be handled roughly.

Knowledge of the enzyme, i.e., its catalytic nature and requirements for optimum performance, is a prerequisite for success in optimizing the inorganic support material. Questions which must be answered with respect to the enzyme are:

1. What are its molecular weight and shape?
2. What is the optimum environment for maximum catalytic activity with respect to: a.) choice of buffer system? b.) feed pH? c.) feed or substrate concentration? d.) activator ions?
3. What is the maximum temperature limit of: a.) the native enzyme? b.) the immobilized enzyme?
4. What ions or other compounds act as inhibitors toward the enzyme?
5. Onto what types of material can the enzyme be bound?
6. What immobilization technique will be employed?
7. How will the IME be used?

Once armed with the answers to these questions, we are in a position to optimize the carrier material for a specific IME application.

Utilizing a phase-separable, sodium borosilicate glass of the following nominal composition:

SiO_2	68.0 by weight
B_2O_3	25.0 by weight
Na_2O	7.0 by weight

Haller (1) and the Russians (2) have shown us how to produce a porous glass with a closely controlled pore morphology. More recently, technology has been developed at Corning by Messing (3) for the production of controlled pore ceramics. Figures 2 and 3 show comparisons of the typical pore morphologies obtained by mercury intrusion techniques for CPG, porous ceramics and other conven-

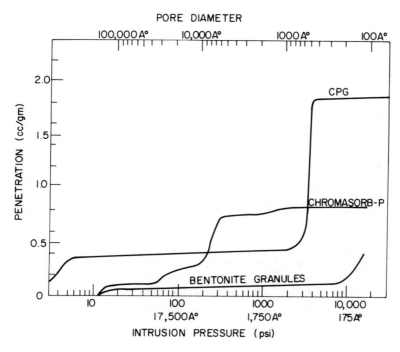

Figure 2: A comparison of controlled-pore glass and other commercial catalyst pore morphologies by mercury intrusion.

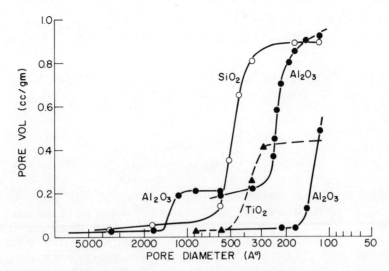

Figure 3: Controlled pore ceramic pore morphologies as measured by mercury intrusion techniques.

tional catalyst materials. Using the CPG process which employs a phase separable, alkali borosilicate glass, and subsequent acid and alkali leaching, one can produce a series of porous glass carriers which range in pore diameter from approximately 30 to 3000 Å. Controlled pore ceramics can be produced using special ceramic technology in selective pore diameter ranges. This ability to selectively produce a porous material with a well-defined morphology is the starting basis for optimizing the inorganic portion of the IME composite.

Of importance in developing or optimizing a support for IME applications are its physical, chemical and mechanical properties. Ideally, a carrier should be chemically inert, have a high surface area (>10 m^2/g), have adequate mechanical strength and cost only pennies per pound. Unfortunately, ideal combinations of these attributes are not always possible and compromises must be made.

To aid in selecting the carrier with the best possible combinations of characteristics for immobilizing biologically active materials, we have developed the decision tree that considers the aspects presented in Fig. 4.

SUPPORT STUDY DECISION TREE

1. What is the pore morphology of the material?
2. Can your enzyme be immobilized on this support?
3. Is it chemically durable in:
   A. Acid environments?
   B. Neutral environments?
   C. Alkali environments?
4. Can the material be conveniently handled?
5. Does it have adequate compressive strength?
6. What is the maximum enzyme loading capability of the system?
7. What pressure drop can be tolerated in the system?
   A. Particle size?
   B. Flow rate?
   C. Shape of particle?
8. What is the operational half life of the system as a function of
   A. Temperature?
   B. pH?
   C. Other conditions?
9. Under what conditions can it conveniently be stored?
10. What is the economic value of the IME system?

FIGURE 4

The first real question we must answer is "What is the best pore morphology with respect to the enzyme's needs?" This selection of pore size is based on two physical considerations: The pore size selected must be large enough to permit easy entry of both the enzyme and the feed substrate into the porous carrier material; and the number of potential coupling sites must be directly related to the available surface area. Surface area is inversely related to the pore size as shown in Fig. 5. Since surface area unaccessible to the enzyme or the feed is useless, one must either estimate or experimentally determine the critical steric diameter of the pore from data regarding the molecular weight and shape of the enzyme and/or the feed. In the case of the enzyme glucoamylase, its molecular weight is approximately 90,000 to 100,000 and it is oblatespheroidal in configuration. Based on this information, one can calculate the radius of gyration of the molecule and estimate a critical exclusion diameter of about 250 Å. Experimentally, this critical diameter

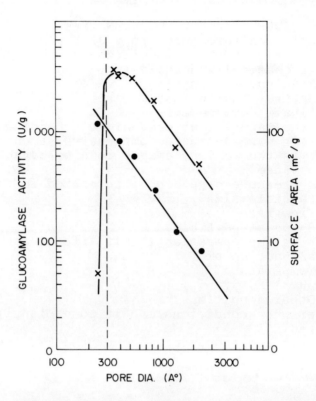

Figure 5: The relationship of pore size to surface area and enzymatic activity is shown. Surface area data is represented by the solid dots while the enzymatic activity results are shown by the X's.

can be obtained by conventional liquid chromatography exclusion studies or enzyme coupling studies of different pore size carriers and the enzyme of choice. If the latter approach is taken, we obtain a curve with a sharp break like that shown by the X's in Fig. 5. When we plot the average pore diameter against enzymatic activity, activity increases as the pore size decreases (or as the surface area increases), until we reach the critical diameter where either the enzyme or the feed molecule is too large to easily enter the porous material. At this point, activity drops off rather markedly. For glucoamylase, we

OPTIMIZATION OF POROUS MATERIALS FOR IME SYSTEMS

observed this drop in activity to occur when pore diameters are in the region of 300 Å. As a consequence, we ended up selecting a material which had a pore diameter slightly above the critical exclusion limit of 300 Å. Our choice was a 350 Å controlled-pore glass with a typical surface area of 90 m^2/g.

Our second consideration in optimizing a carrier for a given biological application is the selection of a chemically inert carrier composition. This is critical both from the viewpoint of the enzyme, and of the product being produced. In the case of the enzyme, corrosion of the support material can either shorten the enzyme half-life by undercutting the base to which it is immobilized, or it can cause deactivation of the catalytic site with

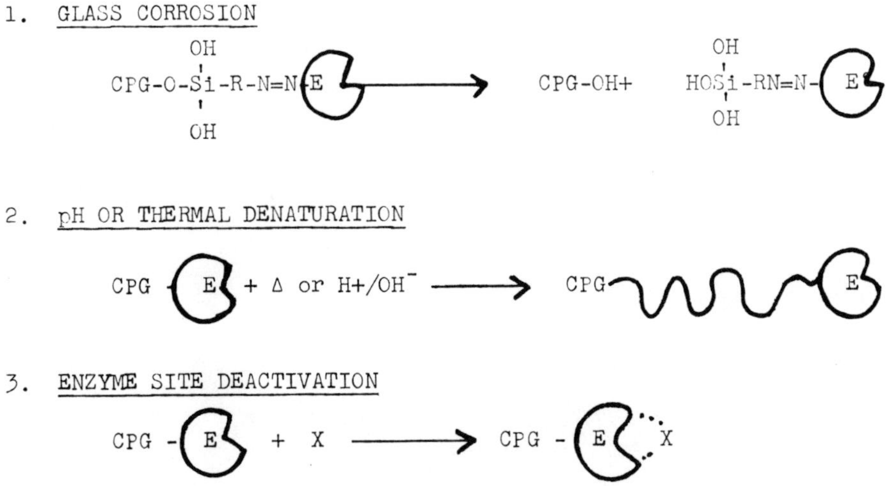

1. GLASS CORROSION

    CPG-O-Si(OH)(OH)-R-N=N-E → CPG-OH + HOSi(OH)(OH)-RN=N-E

2. pH OR THERMAL DENATURATION

    CPG-E + Δ or H+/OH⁻ → CPG~~~~E

3. ENZYME SITE DEACTIVATION

    CPG-E + X → CPG-E⋯X

Figure 6: Three mechanisms whereby enzymatic activity can be lost are illustrated in this figure. Corrosion of the support material can contribute to this loss in activity by undercutting the enzyme bonding site or deactivation of the enzyme's active site.

soluble corrosion products. Both mechanisms are shown in Fig. 6 and lead to a reduction in enzymatic activity. Dissolution of the support material also creates problems in product effluent streams either by increasing product clean-up costs or by violating critical FDA requirements for certain specific ions like mercury, cobalt, lead, etc.

Knowledge of the enzyme's requirements and its preferred environment with respect to buffer systems, operating temperatures, pH and substrate is essential in obtaining a chemically resistant carrier. Also of importance are the conditions under which the IME will be employed. Is the system static or dynamic? Will it be used for a short time or is this a continuing, long-term application?

In the case of glucoamylase we have an enzyme which has a broad pH optimum of approximately 4.0 to 4.5 (Fig. 7). This means we must select an acid-resistant material. Since this particular application involves a long-term usage of the material under dynamic conditions, a

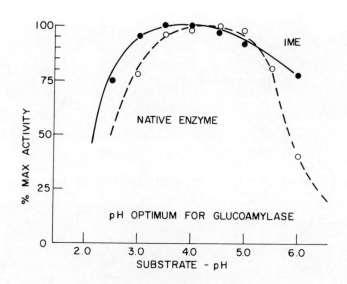

Figure 7: pH profiles for the native enzyme, glucoamylase, and immobilized glucoamylase are shown by the dashed and solid curves, respectively. The pH optimum is shifted slightly to the acid side by immobilization of the enzyme.

DURABILITY TEST RESULTS FOR SUPPORTS

Material Description	STATIC TEST (Mg/M^2-16 hrs.)		DYNAMIC TEST (Mg/M^2-day)		
	1% NaOH	5% HCl	pH 4.5	pH 7.0	pH 8.2
TiO$_2$	0.2	0.8	.05	.05	
ZrO$_2$	0.2	1.1	0.004	0.004	0.01
Al$_2$O$_3$	0.6-0.8	2.0	0.056-0.086	0.01	0.01
Al-Silicate	1.85	3.65	.08-0.1	.02-.05	0.06
CPG-ZrO$_2$ Coat	1.3-2.0	0.2	.03	0.7	0.7-0.9
CPG	3.06	0.08	0.2	0.5	0.3

FIGURE 8

rather severe corrosion problem may exist, even for some of the best materials being considered. The problem is actually compounded because optimum conversion conditions are obtained at temperatures of 40 to 70°C, and a large surface area (>50 m^2/g) is necessary for optimum specific activities.

Initial static screening tests indicated that controlled-pore glass would be an excellent choice (Fig.8). These tests were conducted by immersing a one-gram sample in 10 ml of either a 1% sodium hydroxide or 5% hydrochloric acid solution. The sample was then shaken in a water bath for 16 hours at 60°C. After washing repeatedly in distilled water and then in methyl alcohol, the sample was dried and the weight loss recorded. Subsequent testing under more realistic, dynamic column conditions (Fig. 9), at pH 4.5 with a sodium acetate buffer system, indicated that this was not as wise a choice as originally thought. Fortunately, at about this time a process for applying oxide coatings to glass was developed at Corning. By using this technology, a process was developed for coating controlled-pore glass with zirconium. Essentially this consisted of a vacuum impregnation of either an aqueous or organic solution of zirconium salts. As we can see in Fig. 8, this improved

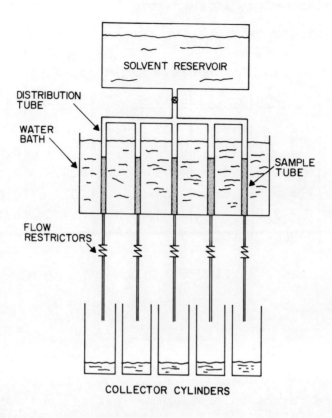

Figure 9: A schematic of the apparatus employed in dynamic durability tests.

the dynamic durability by an order of magnitude. Incidentally, this use of the zirconium oxide film also improved the half-life of the glucoamylase system by an order of magnitude. Half-lives as measured in our labs went from 5 to 50 days. We believe this was a result of our substantially reducing the undercutting of the covalently attached enzyme through the use of the more acid-resistant zirconia film. Optimization of the zirconia coating with respect to alkali durability revealed that

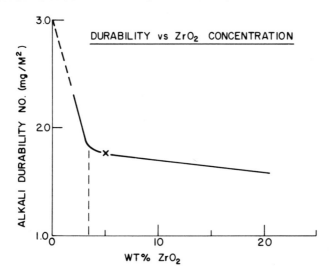

Figure 10: Alkali durability test results showing the minimum weight percent zirconia coating necessary to achieve improved corrosion resistance.

at least a 4%-by-weight ZrO_2 film was necessary to get adequate improvement under dynamic conditions. This can be seen as a break in the curve in Fig. 10.

If one were to consider CPG for other applications, it would be wise to conduct extensive, long-term dynamic tests using it under simulated operating conditions. As we can see from Figs. 11 and 12, the choice of buffer system, pH and dynamic flow conditions have profound influences on the durability and pore morphology of the glass. Even more interesting is the choice of pore and particle size, as shown in Fig. 13. This increase in corrosion rate with increase in particle size can be related to changes in void volume of the column and higher liquid throughputs at given linear velocities. Thus, we see that even rather subtle changes in particle size dis-

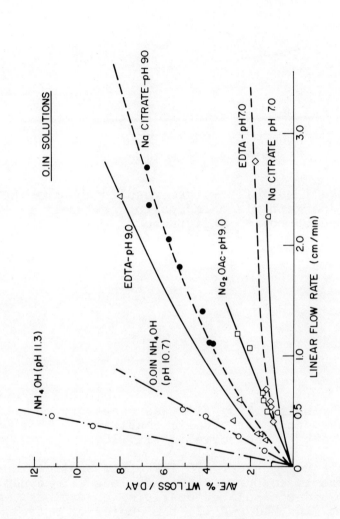

Figure 11: Experimental results showing the effects of flow rate and choice of solvent system on corrosion weight loss. Losses observed are directly related to both flow rate and choice of pH.

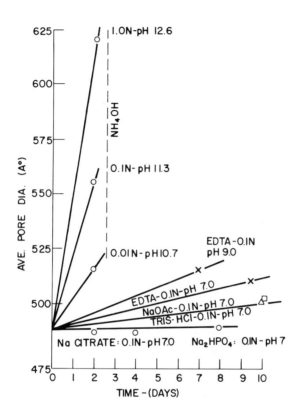

Figure 12: Experimental results showing the change in pore diameter as a function of the time of exposure and choice of solvent system. The more alkaline the pH of the solution, the greater the degree of corrosion.

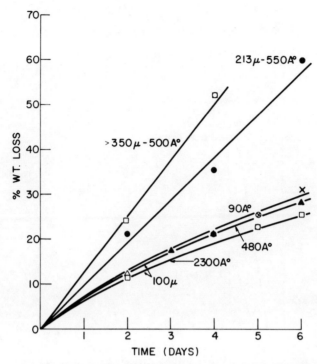

Figure 13: Experimental results showing the influence of pore and particle size on the corrosion of CPG. Particle size has a rather profound effect due to larger packing void volumes and subsequently higher liquid throughputs.

tribution or flow rate may seriously affect the long-term performance of a selected material. Therefore, it is imperative that an exhaustive durability testing program be conducted if satisfactory durability resistance is to be achieved for a given application.

Having optimized the carrier material with respect to enzyme loading or coupling activity, and satisfied ourselves that its chemical durability is adequate for the application at hand, we can now address ourselves to some of the less critical, but still important, questions regarding its physical and mechanical characteristics. One of the first questions that arises is: What particle size distribution and shape shall we employ during scale-up? The answer depends upon determining the maximum allowable pressure drop and linear flow rate for the system. For instance, at a flow rate of 2500 ml per minute for a 100-cm^2 cross-sectional area column, and with column safety release valves designed to vent at 50 psi, we can see from Fig. 14 that we would have to use an 80/120 mesh material or larger. If we decide on the 80/120 mesh material, it might be wise to produce it in a spherical shape which has a lower packing density and a higher void volume fraction than an irregularly shaped particle. This higher void volume would permit higher flow rates at any given pressure or, conversely, would permit lower

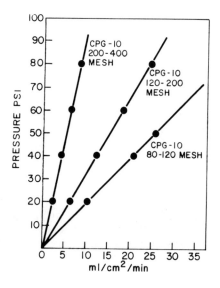

Figure 14: Experimental results for CPG showing the effect of particle size distribution choice on pressure drop and flow rate.

pressures at any given flow rate. Sometimes this is not the final answer. If too large an average particle size is selected for a material with near-critical pore diameter requirements, diffusion restrictions may cause an apparent loss in activity. Should this be the case, one should expect to see a curve similar to the one shown in Fig. 15. Diffusion problems would be suspected if the enzymatic activity increased with a decrease in particle size.

Knowing the maximum allowable operating pressures, hydrostatic heads expected, and the type of system in which the IME will be employed, the questions of whether or not the material has adequate compressive strength and abrasion resistance must now be answered. Neither question is easy to answer from theoretical principles for packed aggregate beds. Therefore, the experimental approach to answering these questions is taken.

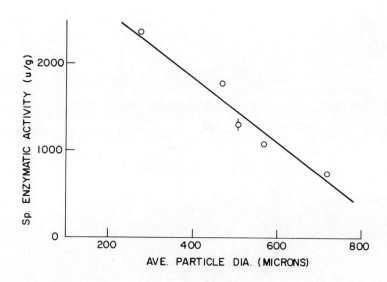

Figure 15: Typical results expected for a diffusion restricted material. Enzyme activity increases as the particle size and diffusion path decrease in size or length.

Using an approach similar to that outlined by Turba (4), we have measured the change in particle size distribution of a well defined powder, such as 30/45 mesh, as a function of applied load to a standard volume of material in a punch and die set. Results from this type of test for CPG have shown that particle breakage is of the order of one to two percent up to 200 psi and about 10 percent at 500 psi. Using this type of information, we can get a good feel for whether we will encounter serious problems later on in scale-up.

As far as attrition testing goes, we have developed wet and dry tumbling tests which measure the percent fines generated from a well defined powder, such as 30/45 mesh. This test involves the use of a fluted, 500-ml flask rotating at approximately 20 rpm for 20 hours. On completion of tumbling, the material is washed free of fines in a fluidized bed, dried and weighed. If weight losses are less than 0.01 g/g of material, we considered this to be a positive indication that the material will perform satisfactorily in most static bed systems. Should these numbers be exceeded, one can anticipate problems involving changing pressure drops or flow rates in continuous flow systems. Neither of these is desirable from an operator's viewpoint.

At this point, one should have satisfied himself that the potential support material has been optimized for:
1. Maximum number of potential coupling sites - highest possible surface area, smallest allowable pore size.
2. Adequate durability, for at least simulated operating conditions.
3. Sufficient strength and hardness so generation of fines will not pose a problem.

Having completed this optimization of physical, chemical and mechanical characteristics of the support, the real optimization of the carrier process must now start. How much can we afford to pay or charge for the carrier materials in terms of dollars per pound of product produced? While I do not intend to answer this question today, I will say that it really depends on the ingenuity of the process chemist, chemical engineer or ceramist in developing a process that uses an inexpensive raw material and as little labor as possible. In this sense, controlled pore ceramics are a major achievement. We hope to reduce support costs by at least an order of magnitude over those currently being charged for controlled pore glasses.

In closing, I would like to acknowledge the work of many of our people at Corning Glass Works. A note of special thanks goes to Mr. Weetall who was instrumental, along with Mr. Tomb, in showing us the significance of using zirconium oxide coatings to improve enzyme half lives and support durability. Also to the chemical engineers of our group, especially Dr. Havewala, for his assistance in measuring many of the physical characteristics so important in developing scale-up data for applications such as the conversion of starch to dextrose, and to Mr. Messing for his foresight and wisdom in the development of controlled pore ceramics.

References

(1) Haller, W., "Rearrangement Kinetics of the Liquid-Liquid Immiscible Microphases in Alkali Borosilicate Melts", J. Chem. Phys. Vol. 42, No. 2, pp 686-693 (1965).

(2) Dementeva, M.I., et al, "Applications of Coursely-Porous Glass in Gas-Liquid Chromatography," Rus.J.Phys. Chem., Vol. 36, No. 1, p. 114 (1962).

(3) Messing, R.A., "A Glucose Oxidase-Catalase System Immobilized in Control Pore Titania Monitored by a Conductiometric Technique," presented at the 166th ACS National Meeting, Chicago, Illinois, 8/27-8/31/73. To be published in the Proceeding of the ACS Immobilized Enzyme Symposium by the Plenum Publishing Corp.

(4) Turba, E., "Behavior of Powders Compacted in a Die," Proc. Brit. Ceram. Soc., No. 3, 101-115 (1965).

WATER ENCAPSULATED ENZYMES IN AN OIL-CONTINUOUS REACTOR-KINETICS AND REACTIVITY

R. I. Leavitt, F. X. Ryan and W. P. Burgess

Mobil Research and Development Corporation

Princeton, New Jersey

The organism used in this study was a species of Pichia yeast. Media and culture conditions have been described previously[1]. Cultures were routinely grown in the presence of 3 weight percent of n-hexadecane as the sole source of carbon. Washed cell suspensions prepared for sonic extraction consisted of approximately 27 grams of wet cells (25% dry weight), suspended in 80 ml. of pH 8.0 phosphate buffer. Sonic extracted enzyme solutions were separated from cell fragments by centrifugation for 15 minutes at 34,000 x g. at 0°C. Generally, the enzyme solutions contained 10-12 mg/ml. of protein.

The enzyme-catalyzed oxidation of n-tetradecyl alcohol to n-tetradecyl aldehyde was examined using the extracts prepared as described. Substrate consisted of a 10 percent (v/v) solution of n-tetradecyl alcohol in n-tetradecane. Since two oxidizable substrates were present in the reaction mixtures, it was necessary to ascertain unambiguously the source of the product(s). C^{14}-1-n-tetradecane and unlabelled n-tetradecyl alcohol were used to prepare a substrate mixture and the enzyme-catalyzed formation of aldehyde from this mixture was examined. Reaction mixtures containing in 20 ml of water:TRIS buffer, pH 8.0, 0.26M; protein, 60 mg; n-tetradecyl alcohol, 404µ mole, and n-tetradecane, 3.8 m mole, were shaken in indented Erlenmeyer flasks for 3 hours at 34°C. Products and unreacted substrate were extracted with 2-propanol-hexane (2:1) and their concentrations and identity were determined using gas liquid chromatography. The primary product detected in this system was n-tetradecyl aldehyde. The aldehyde was isolated from the organic extracts by treatment with aqueous bisulfite reagent. Ether washes of the isolated bisulfite adduct were checked for radioactivity to assure complete removal of any contaminating hydrocarbon. The isolated adduct was broken by extraction with a mixture of sodium hydroxide and ether. The radioactivity of the aldehyde was then determined with a liquid scintillation spectrometer. The specific activity of the aldehyde formed from the alcohol-paraffin charge stock was less than 1.0 percent of that of the paraffin (Table 1) demonstrating that almost all of the aldehyde was derived from the alcohol.

Although n-tetradecyl alcohol is an essentially water-insoluble substrate, its rate of oxidation should bear some relationship to the ratio of hydrocarbon to water in a well-agitated reaction system. This relationship should hold as long as diffusion of hydrocarbon to the cell is rate-limiting. In a series of experiments we found that, indeed, the rate of aldehyde formation increases with increased oil concentration indicating that the rate of alcohol oxidation was limited by the transport rate of the reactants from the oil to the water phase. (Figure 1) The maximum rate observed was approximately 0.6 mg/mg E/hr at an organic/water ratio of 1:2. Further increase of the hydrocarbon-to-water ratio resulted in a phase inversion; i.e., water (enzyme) droplets suspended in a continuous organic phase. In such a system only a single product, n-tetradecyl aldehyde, was detected and an initial rate of aldehyde formation was found to be 1.5 mg/mg E-hr which is significantly higher than that found in the oil-in-water system. The addition to the oil of surfactant

Table 1. Formation of n-Tetradecyl Aldehyde. Formation of n-tetradecyl aldehyde in the presence of labelled n-tetradecane and unlabelled n-tetradecyl alcohol.

^{14}C-1-n-tetradecane- Specific activity (CPM/mmole)	1.5×10^4
n-tetradecyl aldehyde- Specific activity (CPM/mmole)	6.3×10^1

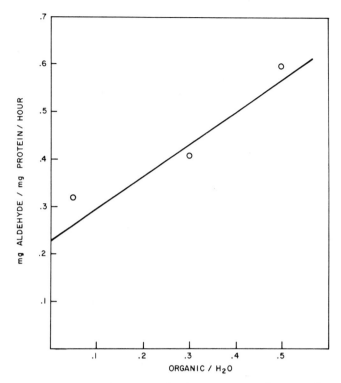

Figure 1 - Effect of Organic/Water Ratio on Reaction Rate

(Igepal 430) resulted in an increase in the initial rate of reaction, reaching a maximum of 2.2 mg/mgE/hr. at a 1% surfactant concentration. At a surfactant concentration above 1%, the reaction rate was reduced (Figure 2).

Kinetic studies which attempt to describe catalyst stability and rate of product formation are difficult to accomplish using the batch systems previously described. In an effort to construct a continuous flow reactor an attempt was made to use an Amicon ultrafiltration cell for our enzyme system. Initially the ultrafiltration membrane designed for the cell was found to be unsuitable since only water could be withdrawn from the reactor with both the enzyme and hydrocarbons being retained within the cell. This was true whether an oil-in-water or water-in-oil system was used. However, we found that the ultrafiltration membrane could be modified by pretreatment to permit only oil and oil-

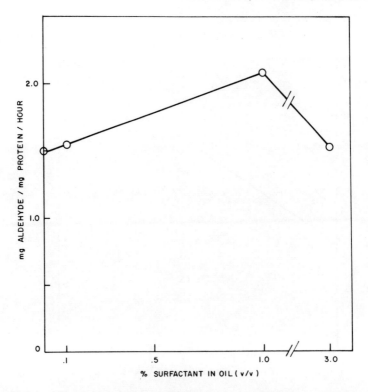

Figure 2 - Effect of Surfactant on Water-in-Oil System

soluble products to pass from the reactor into collecting tubes. Modification of the membrane consisted of washing the membrane (PM 30) in 50% 2-propanol followed by a water wash. The membrane was then autoclaved at 120°C for 15 minutes and dried overnight by sucking air through it. This membrane in combination with a continuous phase oil-water reaction mixture now permitted us to attempt a kinetic study of our enzyme catalyzed hydrocarbon oxidation (Figure 3). The oil-continuous reaction system consisted of 20 ml of water containing 800μ moles TRIS buffer, pH 8.0 and 120 mg of enzyme. The water-enzyme solution was suspended as droplets in 78 ml of a mixture of n-tetradecane and n-tetradecyl alcohol (100 ml/10 g). Droplets were promoted and maintained by a magnetically driven Teflon stirring bar.

The level of charge stock within the reactor was controlled by an electric eye which activated a Milton Roy pump connected to a feed-stock reservoir. Flow

1. Oil Reservoir
2. Pump
3. U F Cell
4. Collecting Vessel

Figure 3 - Continuous Flow Reactor

through the membrane was maintained by pressure (2 psi) at 6.0 ml/hr. Samples (6.0 ml) were collected and monitored for aldehyde production. Flow through the membrane and reactor volume remained essentially constant during the course of the experiment.

The enzyme system had an apparent induction period of about 2 hours (Figure 4) during which time no detectable product was found. After the induction period, the concentration of n-tetradecyl aldehyde, the only product found, increased with on-stream time and reached a maximum concentration of 10 mg/ml oil and then slowly decreased over a period of 55 hours. Using a stirred tank reactor model; (appendix) assuming that the oil was well mixed and that the enzyme containing water volume was constant, it was found that the data were well described by the model with the following parameters.

1) Maximum product formation rate, K_o = 2.08 g/g-enzyme/hr.

2) Enzyme deactivation rate follows an exponential decay function $K_o e^{-\alpha \theta}$, where α = decay constant (hr.$^{-1}$) and θ = on stream time, hr.

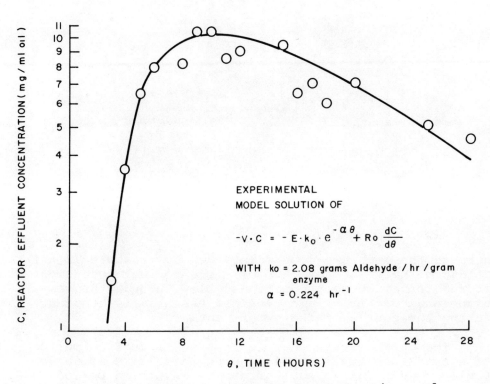

Figure 4 - Reactor Effluent Concentration and Model Solution

3) the half life of the enzyme $T_{1/2} = \ln 2/\alpha$, was found to be 3.1 hours.

The integration of the product over time predicts that a total of 1100 mg of aldehyde will be produced with 120 mg of enzyme - (corresponding to 9.2 gm of aldehyde per gram of enzyme). This is in excellent agreement with the amount actually collected from the reactor effluent. It is also noted that the enzyme was essentially inactive after 17 hours.

Using this model, we calculated the enzyme half life of batch experiments in which Igepal 430 was added. (Table 2) We found that in addition to the increased activity as discussed previously, the addition of Igepal 430 decreased the induction period and increased the half life of the enzyme from 2 to 4 hours (Figure 5).

The increase in enzyme activity in response to the addition of surfactant is believed to be a result of a reduction in the diffusional limitations imposed on the system by the size of the enzyme-water droplets. In the presence of surfactant, the droplet size is reduced, thereby increasing the total surface area available for diffusion of either substrate and oxygen to or products away from the site of catalysis. The explanation for the observed increase in stability of the enzyme system is not immediately obvious. Since there have been reports of protein-surfactant interactions which have resulted in the stabilization of the helical protein structure, it is possible that Igepal 430 and this enzyme could form such an associative complex of greater stability than the native enzyme.

Table 2. Effect of Surfactant on $T_{1/2}$ of Hydrocarbon-oxidizing System

I 430 Volume % in Oil Phase	$T_{1/2}$ Hours
0.00	2.00
0.01	2.57
0.10	3.50
1.00	4.00

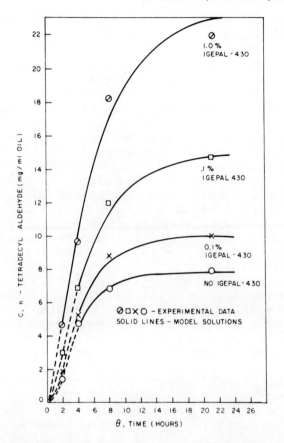

Figure 5 - Effect of Surfactant on Product Formation

Finally, the experimental technique in which an Amicon ultra-filtration cell with a modified oil-permeable membrane was used to study enzyme kinetics should have general applicability to hydrocarbon enzyme systems. Oil continuity provides a gentle and inexpensive means for enzyme encapsulation in addition to optimizing the diffusion of reactants to and oil-soluble products away from the site of catalysis. In combination with a modified ultra-filtration membrane, the water-in-oil reactor represents a unique approach to enzyme engineering which could be extended to include other non-biological processes.

References:

1. Coty, V. F. and Leavitt, R. I. (1971) Microbial Protein from Hydrocarbons. Developments in Industrial Microbiology 12, 61-71.

2. O. Levenspeil, "Chemical Reaction Engineering," John Wiley and Sons (New York, 1962)

Appendix:

The effluent concentration of aldehyde in the oil is described by the ordinary differential equation for a stirred tank reactor (2) (assuming that the oil is well mixed and that the enzyme containing water volume is constant, and a zeroth order reaction)

$$-v_{out} C = -E K_o \exp(-\alpha \theta) + R_o \frac{dC}{d\theta} \quad (1)$$

where θ : time (hrs.)

C : reactor and effluent aldehyde concentration (mg/ml oil)

K_o : product formation rate (gm/hr/grams enzyme)

α : decay constant (hr.$^{-1}$)

$T_{\frac{1}{2}}$: half life of enzyme, $T_{\frac{1}{2}} = 0.693/\alpha$ (hrs.)

$C_{in} = 0$: inlet concentration of aldehyde is zero

$v_{out} = v_{in}$: flow rate, out and into the reactor (6 ml oil/hr)

E : Total enzyme (mg)

R_o : volume of oil in reactor (78 ml oil)

The two unknown parameters in equation 1 are K_o and α (or $T_{\frac{1}{2}}$).

The data was well described by the solution of equation 1 with

K_o = 2.080 gm/hr/grams enzyme

$T_{\frac{1}{2}}$ = 3.1 hours

as displayed in Figure 4. The solution of the differential equation was obtained numerically with a standard integration routine using a 4th order Runge-Kutta method.

In a batch reactor, equation (1) reduced to:

$$0 = -E K_o e^{-\alpha\theta} + R_o \frac{dC}{d\theta} \qquad (2)$$

The solution of equation 2 for C as a function of time is

$$C(\theta) = C_o + \frac{E K_o}{R_o} \frac{T_{\frac{1}{2}}}{\ln 2} \left[1 - e^{-\alpha(\theta-T)} \right] \qquad (3)$$

where T is a fixed delay time (2 hours) and C_o is the aldehyde concentration at time equal to T. Due to an undetermined induction period in the enzyme activity the model equation (3) is assumed to fit the data after this arbitrary delay period of two hours. The parameter $T_{\frac{1}{2}}$ which best describes the experimental data is given in Table 2. The analytical solution is plotted in Figure 5. No attempt is made to calculate an induction period from the data although qualitatively the 1% surfactant data appears to show a reduced induction period.

Acknowledgment:

The assistance of Dr. N. Y. Chen for his helpful consultations during the course of this work is gratefully acknowledged.

ANALYSIS OF REACTIONS CATALYZED BY POLYSACCHARIDE-ENZYME DERIVATIVES IN PACKED BEDS

M. H. Keyes and F. E. Semersky

Owens-Illinois Technical Center

Toledo, Ohio 43666

The reaction catalyzed by an insoluble enzyme derivative in a packed reactor bed is typically analyzed by the integrated Michaelis-Menten equation (1,2,3,4). The value of K_M is found to vary with the flow rate of the substrate solution through the bed. Lilly, Hornby and Crook (1) suggest that this phenomenon is due to the effect of diffusion limitations on the rate of reaction in the case of ficin insolublized to CM-cellulose.

Although it is certainly true that the catalysis of many insoluble enzyme derivatives is limited by diffusion, we present evidence in this paper that derivatives prepared by covalently bonding to CNBr activated Sepharose and agarose appear free from diffusional effects. While collecting this evidence a new method of analysis was developed which uses the Michaelis-Menten equation in the differential form commonly used in soluble enzyme applications. The results derived from this method suggest that this method could be modified to a study of the detailed mechanism of catalysis of immobilized enzymes.

PREPARATION OF INSOLUBLE ENZYMES

Hog liver uricase, Grade II, was obtained from Miles Laboratories, Inc. with an activity of 200 units/gm while jack bean urease, bovine pancreas ribonuclease, and α-chymotrypsin were products of Worthington Biochemical Corporation. The urease was labeled CODE:UR with an activity of 1500 units/mg and the ribonuclease was CODE:RASE with an activity of 300 units/mg. The α-chymotrypsin sample was prepared free of autolytic products (5) and was CODE:CDS.

Four grams of agarose were swelled in 20 ml of distilled water for one hour. In the case of Sepharose 4B, the material is obtained in aqueous suspension and an equivalent amount of Sepharose was washed in 20 ml of distilled water. Ten grams of cyanogen bromide in 60 ml of distilled water were added to the polysaccharide suspension. The pH was adjusted and maintained at 10.5 by addition of 3.0 M NaOH. After activation, the derivative was washed with an excess of 0.05 M borate buffer at pH 8.5.

The activated Sepharose was used to insolubilize urease. First the Sepharose derivative was washed with 300 ml of 10^{-3} M HCl. Next a solution of urease was prepared by the addition of 20 mg of urease to 7 ml of a solution that contained 0.1 $NaHCO_3$ which also contained 0.5 M NaCl. The insoluble derivatives of ribonuclease, uricase, and chymotrypsin were prepared in the same way using agarose described above and all derivatives were stored at 3 ± 2°C.

The freshly prepared immobilized ribonuclease contained activity equivalent to 1.4 mg of soluble ribonuclease per ml of swelled support. Urease, chymotrypsin, and uricase contained 3.7, 0.4 and 0.05 mg/ml respectively. No attempt was made to measure the actual quantity of bound protein.

MEASUREMENT OF REACTION RATES

The rate of reaction of cytidine-2-3 cyclic monophosphate in the presence of the insoluble ribonuclease derivative was measured by two methods: batch method and column method. Other insoluble enzyme derivatives were analyzed by the column method only. In both schemes most of the data were obtained by a spectrophotometric measurement of the rate of disappearance of substrate or appearance of product catalyzed by an insolubilized enzyme. In our laboratory the spectral detector was either a DU-2 or DK-2A Beckman spectrophotometer. The substrate solutions for the ribonuclease (6) and uricase (7) derivatives were both monitored by the change in absorbance at 290 nm while the wavelength of 256 nm was selected for chymotrypsin (8).

The hydrolysis of urea to ammonia and carbon dioxide catalyzed by the urease derivative was monitored by the change in potential of a cation electrode caused by the formation of ammonium ions in the urea solution. Experiments with urease required collection of eluent and measurement in intervals rather than continuous flow detection.

Batch Method

In the batch method, a substrate solution was stirred with a variable speed magnetic laboratory stirrer. A stock enzyme suspension was added to the substrate solution at time zero. A fraction of suspension was withdrawn and filtered with 0.2 to 3μ filters and the absorbance of this aliquot recorded along with the time of withdrawal. Usually several samples were taken during the first two to five minutes of the reaction and measured spectrophotometrically.

Column Method

The column method requires packing a glass column (2.0 or 2.8 mm x 7.5 cm) with insolubilized enzyme to form a packed bed reactor. After packing the column, it was inserted in the flow scheme shown in Figure 1. Thermistors were located on the column and within the sample compartment of the spectrophotometer to monitor the temperature of the reaction and the temperature of measurement.

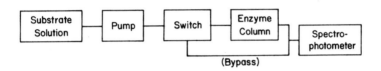

Figure 1. Diagram of the system used in the column method which is the same for all enzymes except urease in which the spectrophotometric detector was replaced by a cation electrode.

A typical procedure involved the following steps. The bypass-channel and packed bed reactor were washed with the same buffer solution to be used when measuring the rate of reaction until no absorbance was detected at the wavelength to be used. Next the substrate solution was pumped through the bypass until a constant absorbance was recorded. The direction of flow was changed so that solution flowed through the packed bed reactor and when constant absorbance was obtained, its value and the flow rate were recorded. The flow rate was determined by measuring the volume of solution which passed through the system per unit time. The above procedure was repeated for each alteration in flow rate until sufficient data points (usually four to ten) were collected to determine the initial velocity.

RESULTS AND DISCUSSION

Theory

Insolubilized enzyme assays can be performed in packed bed reactors by passing substrate solution through the reactor bed at a constant flow rate. Typically, analysis of the data is accomplished by the use of the integrated Michaelis-Menten equation (1,2,3,4):

$$P[S_o] = K_M \ln(1-P) + k_3(E/Q)(V_\ell/V_T)$$

P is defined as the fraction of reacted substrate and Q is the flow rate through the column. The volume, V_ℓ, is the void volume of the column while V_T is the total bed volume.

Plotting log (1-P) versus $P[S_o]$ at constant flow rate typically yields a straight line. Unfortunately the slope of this line which should equal K_M varies with the flow rate of the experiment in which the data were obtained (2). The observed variation in K_M could be due to the reaction rate being diffusionally controlled.

Another factor in these measurements may contribute to the variation in K_M observed. The value of P varies from 0.2 to 0.8 in many cases (1,2,3). The steady-state approximation used in derivation of the Michaelis-Menten equation usually applies only when the rate is measured in the early stage of the reaction ($P \geq 0.3$) (9). Thus the integrated equation is being used in a region of the reaction in which the applicability of the Michaelis-Menten equation is in question.

We analyzed the reactions catalyzed by packed bed reactors using the differential Michaelis-Menten equation. The initial velocities and corresponding initial substrate concentrations to use in this equation can be obtained from experiments with packed bed reactors by plotting the change in concentration of the substrate solution in the eluent versus the residency time of the substrate solution within the column. This residency time could be calculated by dividing the volume of the column by the flow rate. Unfortunately, when the initial substrate concentration was changed, the flow rate necessary to measure the rate of reaction changed appreciably and concomitantly the change in pressure sometimes caused a change in the volume of the column. Therefore, rather than calculate the volume of the column for each experiment, the initial velocities in units of moles per minute were conveniently calculated by simply plotting the change in substrate concentration versus the reciprocal of the flow rate. Since for each experiment, the volume of the column is constant in the linear region used to calculate the initial velocity, no error in the kinetic analysis should arise due to changes in the volume of the reactor bed.

A modification of the Michaelis-Menten equation is necessary to use initial velocities expressed in moles per minute. A similar modification has been discussed for enzymes immobilized in membranes (10).

The enzyme concentration is replace by:

$$[E] = E_T/v$$

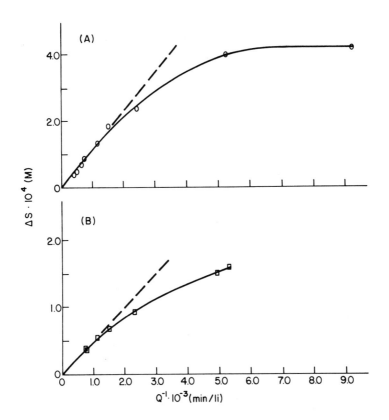

Figure 2. (A) The change in benzoyl tyrosine ethyl ester concentration is plotted versus the reciprocal of the flow rate for insoluble chymotrypsin derivative. The substrate solution was prepared by mixing 140 ml of $1.07 \cdot 10^{-3}$ M benzoyl tyrosine ethyl ester in 50% (w/w) methanol with 160 ml of 0.08 M TRIS, pH 7.8; 0.1 M $CaCl_2$. The temperature was $25 \pm 2°C$. (B) The change in cytidine 2',3' cyclic monophosphate concentration is plotted versus the reciprocal of the flow rate for the insoluble ribonuclease derivative. The initial concentration of cytidine 2',3' cyclic monophosphate was $3.81 \cdot 10^{-4}$ M in 0.05 M sodium acetate at pH 5.0.

where E_T is the total enzyme contained with the packed bed reactor
v is the liquid volume of the column.

The velocity can be expressed as:

$$V = \frac{\partial [S]}{\partial t} = \frac{d(S/v)}{dt} = \frac{\partial S}{\partial t}\left(\frac{1}{v}\right)$$

Making these substitutions in the Michaelis-Menten equation, one obtains:

$$V' \equiv \frac{dS}{dt} = \frac{k_3 E_T [S]}{K_M + [S]} \quad ; \quad \frac{1}{V'} = \left(\frac{K_M}{V_m'}\right)\frac{1}{[S]} + \frac{1}{V_m'}$$

The initial velocity, V', was obtained by plotting the change in substrate concentration in the eluent versus the reciprocal flow rate (Figure 2). For each insoluble enzyme derivative, the reciprocal of this initial velocity was plotted versus the reciprocal of the initial substrate concentration resulting in a straight line as predicted by the last equation shown above.

INSOLUBLE DERIVATIVES

The measurements of the initial velocity for the hydrolysis of cytidine-2',3' cyclic monophosphate by the insoluble ribonuclease derivative required prior dilution of the derivative with swelled agarose. This dilution resulted in 4.31×10^{-3} (v/v) in the final material used to pack the column.

The volume of the insoluble enzyme derivative in the column was 0.28 cc. Experiments performed at several substrate concentrations yielded the data represented in the reciprocal plot in Figure 3. These data were treated by nonlinear least squares analysis using the method described by Cleland (11) to yield a value of K_M of $(4.5 \pm 1.1) \times 10^{-4}$ M and a value of V_m' of $(4.1 \pm 0.6) \times 10^{-8}$ moles/min. Michaelis-Menten constants were also calculated using data obtained from the batch method (Figure 4). The slower stirring rate was the slowest rate that would keep the insoluble ribonuclease derivative suspended while the faster rate was the maximum which could be used without splashing. Note that there is no effect due to changes in stirring rate.

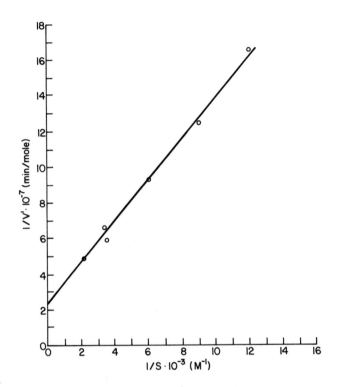

Figure 3. Lineweaver-Burk plot for the insoluble ribonuclease derivative obtained by the column method. The solution contained 0.05 M acetate at pH 5.0 and was maintained at 25 ± 2°C.

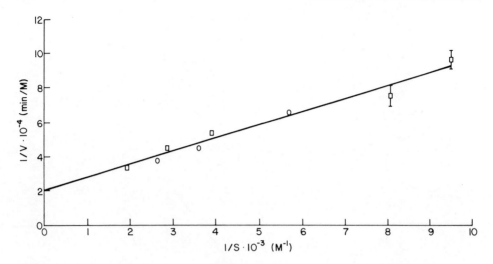

Figure 4. Lineweaver-Burk plot for insoluble ribonuclease derivative obtained by the batch method with cytidine 2',3' cyclic monophosphate as the substrate. The same conditions were used as described for Figure 3. The results of experiments at the slow stirring rate are represented by (o) while (□) represents the results at the faster rate.

Although no estimation of the quantity of ribonuclease insolubilized was made, it was obvious that no more than the total ribonuclease added could have been insolubilized on swelled agarose. Making the assumption that all of the ribonuclease molecules in contact with the cyanogen bromide-treated Sepharose were insolubilized, the value of k_{+2} was calculated for the insoluble enzyme to be 120 min^{-1} by the data obtained in the batch method and 45 min^{-1} for the column method. The change to a lower value in the analysis of data obtained by the column method may be due to loss of active enzyme since one month had elapsed before the column method of assay was applied. The value of k_{+2} for the soluble enzyme is 63 ± 6 min^{-1}. Since the actual quantity of ribonuclease insolubilized is probably considerably less than the total available, k_{+2} must be larger for the insoluble ribonuclease derivative than for the native enzyme.

The increase in the value of k_{+2} can easily be due to changes in the environment of the enzyme upon insolubilization. Figure 5 shows the reciprocal plot yielded by experiments with native ribonuclease in 20 per cent dextran. The value of k_{+2} from this analysis is $(2.7 \pm .5) \times 10^{+3}$ min^{-1} while the value of K_M remains the same as it was in the absence of dextrin. This value of k_{+2} is much higher than those obtained as minimum values for insoluble ribonuclease derivative.

Table I

Ribonuclease

Preparation	Conditions[a]	K_M (M·10^{+4})	k_{+2} (min^{-1})
Native	0.2 M Sodium Acetate[12] pH 5.0, 25°C	4.0 ± 0.3	63 ± 6
	0.05 M TRIS 0.05 M Sodium Acetate[6] 0.1 M KCl pH 5.5, 25°C	5.1 ± 0.7	(1.9 ± .2) x 10^{+2}
Native	20 Per Cent Dextran	5.2 ± 1.7	(2.7 ± 0.5) x 10^{+3}
Insoluble Ribonuclease Derivative	Suspension	5.5 ± 1.7	< 120
Insoluble Ribonuclease Derivative	Column	4.5 ± 1.1	< 45

[a] Unless stated otherwise, all experiments performed in 0.05 M Sodium Acetate, pH 5.0, 23°C.

Table II

Uricase

Enzyme	K_m(M)·10^{+5}	V_m(M/min)·10^{+6}	Borate (M)	pH	T (°C)
Native[7]	2.0	----	> 0.01	7-10	20
CNBr-Agarose 2 weeks	1.8 ± 0.5	8.8 ± 1.1	0.1	8.5	25
CNBr-Agarose 7 weeks	0.25 ± 0.07	1.28 + 0.06	0.01	8.5	25

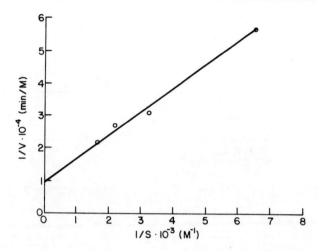

Figure 5. Lineweaver-Burk plot for soluble ribonuclease in 20% (w/w) dextrin. The same conditions were used as described for Figure 3.

Table I lists the Michaelis-Menten parameters for ribonuclease. As can be readily seen in Table I, the values of K_M obtained are all within error of the native ribonuclease. Furthermore, the unexpectedly high minimum value of k_{+2} obtained for the insoluble ribonuclease derivative can easily be explained by the mixed polysaccharide-water environment. The agreement of the value of K_M between the batch and column method (Table II) to the value for the soluble enzyme (6,12) as well as the lack of any effect of changes in stirring rate on the rate of reaction supports the conclusion that there are no diffusional limitations in the rate of reaction.

Michaelis-Menten constants were also obtained for two other insoluble enzymes packed in glass columns which agree quite well with the values for the soluble counterpart. Urease insolubilized to cyanogen bromide activated Sepharose was analyzed in 0.01 M triethanolamine at pH 6.7. Figure 6 shows the data plotted for these experiments. The value of K_M obtained was $(2.8 \pm 1.1) \times 10^{-3}$ M (13).

When uricase insolubilized in agarose activated with CNBr was analyzed by this method, the initial results again showed agreement between the kinetic mechanism for the insoluble enzyme and that of its soluble counterpart (Table II). Seven weeks after preparation, insoluble uricase showed three kinetic differences from the freshly prepared insoluble derivative or native uricase (Figure 7): 1) lower value of K_M; 2) lower value of V_M; 3) a

REACTIONS CATALYZED BY POLYSACCHARIDE-ENZYME DERIVATIVES 279

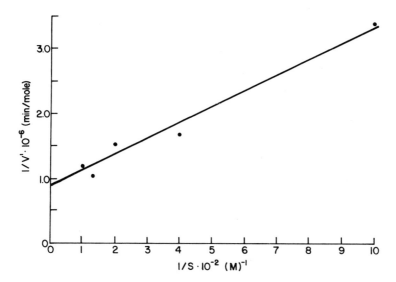

Figure 6. Lineweaver-Burk plot for insoluble urease derivative obtained by the column method with urea. The urea solution contained 0.01 M triethanolamine at pH 6.7 and was maintained at 25 ± 2°C.

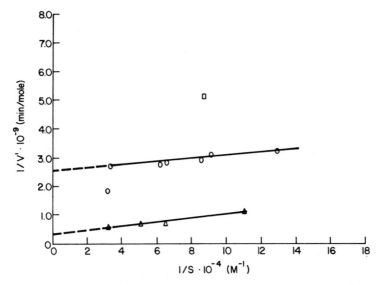

Figure 7. Lineweaver-Burk plots for insoluble uricase derivative obtained by the column method with uric acid. All experiments performed at pH 8.5 and 25 ± 2°C. (□) indicates 0.1 M borate buffer while (o) indicates 0.01 M borate buffer for experiments performed six weeks after preparation of the uricase derivative. (Δ) indicates experiments performed within two weeks after preparation of the uricase derivative in 0.1 M borate buffer.

larger effect of borate buffer concentration on the enzyme activity (7). The effect of high borate buffer concentration could not be explored since the single experiment in 0.1 M borate indicated a very low activity for the insoluble derivative. Apparently, the insoluble enzyme after seven weeks of storage at 4°C no longer catalyzes the reaction of uric acid by the same mechanism as native uricase or a freshly prepared insoluble uricase derivative.

Initially, the three enzymes that were immobilized and analyzed by the differential Michaelis-Menten equation all showed the same values for K_M as the respective native enzymes under the similar experimental conditions. Second, the kinetic parameters obtained by the batch method agree with those obtained by the column method. Finally, the stirring rate used in experiments to determine activity by the batch method did not affect value of the kinetic parameters. Thus it appears that the following conclusions can be formulated:

1. Enzymes insolubilized to porous polysaccharides activated with CNBr usually form derivatives which catalyze reactions by a nondiffusion mechanism.

2. The values of the kinetic parameters can be obtained for these derivatives in packed bed reactors by the use of the differential Michaelis-Menten equation.

In addition, it would seem possible to study insoluble enzyme derivatives in a packed bed reactor by determining the initial velocity as discussed here, but evaluating the data by more complex kinetic schemes. These detailed schemes are probably necessary for a detailed understanding of the mechanism of catalysis for insoluble enzyme derivatives as they have been found to be for their soluble counterparts. Detailed schemes derived for insoluble enzyme derivatives might well lead to more knowledge concerning enzyme catalysis in general, since it would be possible to use the same enzyme sample for experiments that span several weeks since many insoluble derivatives have been shown to be stable with time. Variation between batches of a purified enzyme either due to inherent enzyme differences in treament or during purification could be eliminated.

SUMMARY

Reactions catalyzed by immobilized enzymes in packed bed reactors were analyzed using the differential Michaelis-Menten equation. The values of K_M were calculated for insoluble ribonuclease [$(4.5 \pm 1.1) \times 10^{-4}$ M], uricase [$(1.8 \pm 0.5) \times 10^{-5}$ M], and urease [$(2.8 \pm 1.1) \times 10^{-3}$ M]. These values are within experimental error of those for the solution enzymes under the same experimental conditions. The method of analysis is discussed and would appear applicable to any immobilized enzyme derivatives in which diffusional effects are not observed.

REFERENCES

1. Lilly, M. D., Hornby, W. E., and Crook, W. M., Biochem. J. 100, 718 (1966).

2. Sharp, A. K., Kay, G., and Lilly, M. D., Biotechnology and Bioengineering 11, 363 (1969).

3. Lilly, M. D., and Sharp, A. K., Institution of Chem. Eng. Trans. 215, CE 12 (1968).

4. Bar-Eli, A., and Katchalski, E., J. Biol. Chem. 238, 1690 (1963).

5. Yapel, A., Han, M., Lumry, R., Rosenberg, A., and Shiao, D. F., J. Am. Chem. Soc. 88, 2573 (1966).

6. Hammes, G. G., and Walz, F. G. Jr., Biochim. Biophys. Acta 198, 604 (1970).

7. Baum, H., Hübscher, G., and Mahler, H. R., Biochim. Biophys. 22, 514 (1956).

8. Hummel, B. C. W., Can. J. Biochem. Physiol. 37, 1393 (1959).

9. Walter, C., in Steady-State Applications in Enzyme Kinetics, p. 34. The Ronald Press Company, New York, 1965.

10. Sundaram, P. V., Tweedale, A., and Laidler, K. J., Can. J. Chem. 48, 1498 (1970).

11. Cleland, W. W., Adv. in Enzymology 29, 1 (1967).

12. Herries, D. G., Mathias, A. P., and Rabin, B. R., Biochem.J. 85, 127 (1962).

13. Kistinkowsky, G. B., and Shaw, W. H. R., J. Am. Chem. Soc. 75, 2751 (1953).

THE PREPARATION OF MICROENVIRONMENTS FOR BOUND ENZYMES BY SOLID

PHASE PEPTIDE SYNTHESIS

James B. Taylor and Harold E. Swaisgood

North Carolina State University

Dept. of Food Science, Raleigh, N. C.   27607

It has become evident that in the area of immobilized enzymes no single attachment system or resin support or reactor design will be optimum for all enzymes. Thus it is imperative that we develop a repertoire in these areas to allow for system designs which will either optimize the immobilized enzyme or make it more substrate selective. We sought to develop a systematic method that would allow one to specify the chemical microenvironment on the support resin onto which the enzyme is attached; thereby enabling a more systematic investigation of the influence of surface environment on enzyme structure and activity. Hornby et al., (1) Goldstein et al., (2), and Goldstein (3) have shown that the surface characteristics of the resin, i.e., its charge and hydrophobic nature, do influence the apparent kinetic parameters of immobilized enzymes. Furthermore, we (4) have shown that the spatial placement of the enzyme from the resin's surface also affects the kinetic parameters. This led us to examine the feasibility of synthesizing peptide chains of a given chemical characteristic on the surface of a resin and attaching the enzyme to these chains. Such a method would allow comparisons of microenvironments without the interferring effects of matrix structure or enzyme distance from the surface, since these could be held constant.

For the support matrix we chose porous glass beads with a 700 $\overset{\circ}{A}$ pore size and a 120-200 mesh size. This support has the advantages of structural rigidity, relative chemical inertness and freedom from solvent swelling effects which prevail in organic resins. These characteristics allow for the use of column procedures. Furthermore, the glass surface can be modified by silanization (5) to yield a covalently attached amino alkyl silane. The

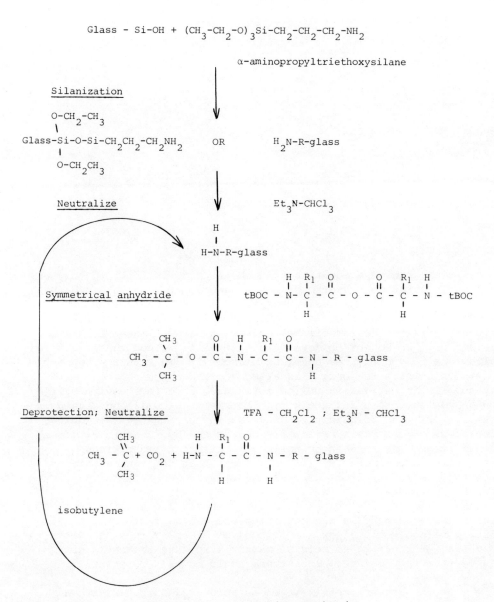

Fig. 1. Reactions in synthesis

amino function was used as the base upon which the desired peptide chains were built. The silanization method used requires that an aqueous 10% solution of α-aminopropyltriethoxysilane be reacted with the glass beads for three hrs (70°C and pH 4) at which time the solution was filtered off and the beads heated overnight at 125°C. The glass beads were washed with acetone then with water to remove excess silane. While we used glass for the support resin, other inorganic carriers (6) such as nickel oxide and other metallic oxides can be used for silanization. Furthermore, one is not necessarily limited to the silane method if a covalently attached amino group can be obtained by another method. Thus the amino groups on Inman's and Hornby's (7) nylon tubular reactors might be amenable to chain synthesis. The primary concern is that the resin have a covalently attached amino group and be compatible with the solvents used in solid phase peptide synthesis.

The essential chemical scheme for chain synthesis is given in Figure 1. Amino acids are added in a cyclic stepwise fashion until the desired sequence is synthesized. The method of synthesis developed in this study differs from Merrifield's (8) classical solid phase peptide synthesis in several respects. First, the initial amino acid is bound by a peptide linkage instead of the usual benzyl ester linkage. This allows for cleavage of side chain protecting groups without cleavage of the chain. Since this procedure prevents any purification of the synthesized chains, all reactions must go to completion to obtain a homogeneous chain population. A second modification is the use of $TFA-CH_2Cl_2$ (50% V/V) to cleave the tBOC amino protecting group. Karlsson et al., (9) have shown that this reagent removed 97% of the tBOC groups in five minutes and 100% in ten minutes. Further, it does not possess the disadvantages of HCl-HoAC which may result in an acetylation of the chain or the formation of peroxide in HCl-dioxane which causes oxidation of certain residues. However, Ragnarsson et al., (10) have reported that the ε-benzyloxycarbonyl protecting group of lysine is slowly removed (∼ 3% in 18 hrs) by $TFA-CH_2Cl_2$. Grahl-Nielsen and Tritsch (11) have found similar results when HCl-HoAC was used to deprotect. To prevent cleavage of the ε-lysine protecting group we used a 20% solution of mercaptoethane sulfonic acid in glacial acetic acid which, according to Loffett and Dremier (12), is selective for the tBOC group and does not cleave the benzyloxycarbonyl group. The third modification was the use of symmetrical anhydrides (13) in place of carbodiimide activation of the carboxyl group. The anhydrides posses the following advantages: 1) they do not undergo a rearrangement like the active O-acyl to inactive N-acyl shift of the carbodiimide-amino acid intermediate, 2) unlike the carbodiimide method the reaction vessel is not subjected to the byproduct formation of an insoluble disubstituted urea, and 3) the unused anhydrides can be recovered, thus allowing for the use of higher concentrations of reactive amino acids. The

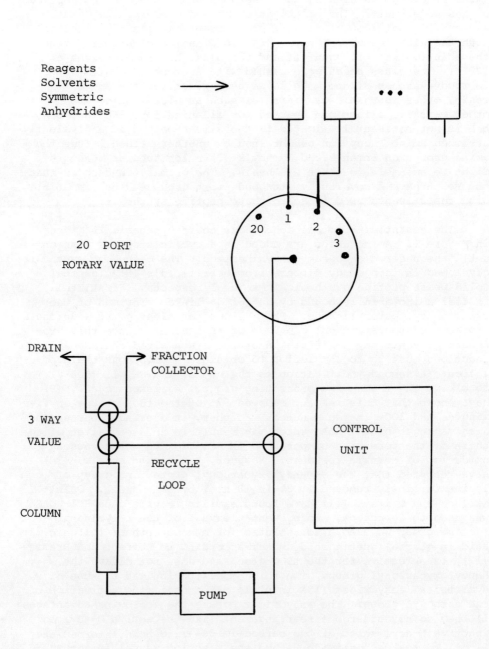

Fig. 2. Automated synthesizer.

collected anhydrides are hydrolyzed back to the amino acid form by reaction overnight in methylene chloride saturated with water. The fourth modification is the measurement of the extent of each coupling step by Dorman's (14) method. In conjunction with this test amino acid analyses were performed on the final product to further confirm the extent of the coupling reaction and chain site homogeneity.

The introduction into solid phase synthesis of a non swelling resin and a coupling reaction of comparable rate to the carbodiimide method but free of the urea formation allowed for the development of an automatic column peptide synthesizer. A column method is preferred over a batch method for two reasons. First, in the case of glass beads, batch methods with their stirring cause considerable attrition. Second, reactions are more efficient in columns since each volume element of the matrix is subjected to a higher concentration of reactants than possible in a batch method with the same time and column parameters. Furthermore, the use of a column procedure permitted the development of a less sophisticated control unit since precise volume and the time accuracy are not necessary. In Figure 2 is shown a simplified schematic of the automated system. It should be noted that the column is used in an ascending flow mode. This facilitates the rapid self purging of any air bubbles introduced into the column. While the system is simple in design the prototype was of a fairly complex nature to allow for flexibility and multipurpose use. Thus the prototype is capable of 15 steps per cycle, nine amino acid residue additions per run and on any given step providing for one of three times, one of three temperatures, collection of effluent, and recycling. The control unit is based on the use of three single pole multiposition stepping relays. One relay is used for sequencing the amino acids. The other two are in tandem with one carrying AC voltage for the timers and the other carrying DC voltage for the switches of the other functions. The DC voltage permits the use of diode arrays to electronically isolate the separate functions. Easy program changes are sustained by controlling all functions through switches.

The sequence of steps and times for a cycle is given in Fig. 3. It takes approximately two hours per residue. While we used a coupling time of 30 minutes this is probably not the shortest time for a 100% yield. Recent reports have shown six minute coupling times for the carbodiimide method (15) and 20 minute coupling times for the anhydride method (13). Also, theoretical considerations by Rony (16) have shown that as the concentration of activated amino acids is increased the time for a 100% yield is dramatically decreased. Furthermore, increases in flow rate would allow for shorter wash times and an overall shortening of the time per cycle. One caveat that should be noted is that the anhydrides must be kept

Fig. 3. Steps in one cycle.

Step	Reagent	Time (min.)	Reagent Vol. (ml)	Collect
1	CH_2Cl_2	6	18	--
2	TFA-CH_2Cl_2‡	10*	15	--
3	CH_2Cl_2	6	18	--
4	$CHCl_3$	6	18	--
5	Et_3N-$CHCl_3$	6	18	--
6	$CHCl_3$	6	18	--
7	CH_2Cl_2	6	18	Yes
8	Sym. Anhyd.	30*	15	Yes
9	CH_2Cl_2	6	18	Yes
10	Pyr-HCl	6	18	--
11	CH_2Cl_2	6	18	--
12	DMF	6	18	--
13	Et_3N-DMF	6	18	Yes
14	DMF	6	18	Yes

*Recycle after 5 min.

‡Used MESNA for lys.

Fig. 4. Chains synthesized.

Chain Type	H Ø/chain* (Kcal/mole)	H Ø/res.* (Kcal/mole)	Charge	No. Res.
$(gly)_3$	0	0	0	3
$(gly)_6$	0	0	0	6
gly-Ala-gly	0.75	0.25	0	3
gly-val-gly	1.70	0.57	0	3
Ala-Val-Ala	3.20	1.07	0	3
Ala-Pro-Ala	4.10	1.37	0	3
Pro-Val-Ala	5.05	1.68	0	3
Phe-Ala-Phe	6.05	2.02	0	3
Pro-Phe-Pro	7.85	2.62	0	3
$(Gly)_3$-Pro-Phe-Pro	7.85	1.31	0	6
$(Pro-Phe-Pro)_2$	15.70	2.62	0	6
Val-Lys-gly	3.20	1.07	+1	3
Asp-Lys-Val	3.20	1.07	0	3
Pro-Asp-Ala	3.35	1.12	-1	3
$(Gly)_2$-Asp	0	0	-1	3
Gly-$(Asp)_2$	0	0	-2	3
$(Asp)_3$	0	0	-3	3
$(Gly)_3$-$(Asp)_3$	0	0	-3	6
$(Gly)_2$-$(Asp)_4$	0	0	-4	6
Gly-$(Asp)_5$	0	0	-5	6
$(Asp)_6$	0	0	-6	6
$(Val)_2$Lys	4.90	1.63	+1	3
Val-$(Lys)_2$	4.70	1.57	+2	3
$(Lys)_3$	4.50	1.50	+3	3

*Based on Bigelow's values.

ABBREVIATIONS: CH_2Cl_2, methylene chloride; $CHCl_3$, chloroform; DMF, dimethylformamide; Et_3N, triethylamine; HOAC, acetic acid; MESNA, mercapto ethane sulfonic acid; Pyr·HCl, pyridine hydrochloride; t-Boc, tert-butyloxycarbonyl; TFA, trifluoroacetic acid.

under strict anhydrous conditions for even a relatively low percent of water in the solution will drastically reduce the yield by hydrolyzing the anhydride to its acid form. To insure anhydrous methylene chloride we routinely distilled predried methylene chloride from phosphorous pentoxide (17) into a holding reservoir containing activated 4A molecular sieves. In order to simplify synthesis of the anhydrides and maintain anhydrous conditions we used jacketed fritted glass disc funnels sealed with drying tubes; these funnels were connected directly to the 20 port value. Since the insoluble disubstituted urea byproduct floats in the solution, direct pumping of the freshly made anhydride from the fritted glass disc funnels is possible. Thus exposure to the air is minimized and preparation of the anhydride is simplified, making the process almost as convenient as the carbodiimide method. For a typical run we used 1 g. of the 700 $\overset{\circ}{A}$ glass beads, 0.1 M solution of each anhydride in methylene chloride, and a reaction temperature of 25°C. The system has a volume of approximately 6 ml and a holdback (18), H = 0.02. The average flow rate was approximagely 180 ml/hr which gave at least two volume changes of the system during washing steps.

The chain types synthesized are shown in Figure 4. Calculated hydrophobicities (19) were used to quantitatively characterize the degree of each chain's hydrophobic nature. The glycine chains of three and six residue length are used as the unperturbed system which permits a better kinetic comparison of the effects of hydrophobicity and charge than the use of native enzyme in solution. Comparison of the chains $(Gly)_3$-Pro-Phe-Pro and (Pro-Phe-Pro) should indicate whether the hydrophobic effect is a function of total or per residue hydrophobicity. We have also synthesized chains with constant hydrophobicity but with varying charge. Finally, chains of various charge magnitudes have been synthesized to study the effect of pH shift and the interaction of electrostatic forces between substrate and matrix. After synthesis the side chain protecting groups were cleaved in HBr-TFA. Before this cleavage the terminal amino group was converted to a carboxyl group by succinic anhydride except for the aspartic acid chains where the amino function was not modified. Thus, either the amino or carboxyl groups on the enzyme, depending on chain type, can be used to immobilize the enzyme. Properties of trypsin coupled to the matrix via these chains are currently being investigated.

REFERENCES

1. Hornby, W. E., Lilly, M. D. and Crook, E. M., (1968) *Biochem. J. 107*, 669.

2. Goldstein, L., Levin, Y., and Katchalski, E., (1964) *Biochem. 3*, 1913.

3. Goldstein, L. (1972), *Biochem. 11*, 4072.

4. Taylor, J. B. and Swaisgood, H. E. (1972) *Biochim. Biophys. Acta, 284*, 268.

5. Weetall, H. H. and Havewala, N. B. (1972) *Biotechnol. Bioeng. Symp. No. 3*, 241.

6. Weetall, H. H. and Hersh, L. S. (1970) *Biochim. Biophys. Acta, 206*, 54.

7. Inman, D. J. and Hornby, W. E. (1972) *Biochem. J. 129*, 255.

8. Stewart, J. M. and Young, J. D. (1969) Solid phase peptide synthesis, W. H. Freeman and Co., San Francisco.

9. Karlsson, S., Lindeberg, G., Porath, J. and Ragnarsson, V. (1970) *Acta Chem. Scand. 24*, 1010.

10. Ragnarsson, V., Karlsson, S. and Lindeberg, G. (1970) *Acta Chem. Scand. 24*, 2821.

11. Grahl-Nielsen, O. and Tritsch, G. L. (1969) *Biochem. 8*, 187.

12. Loffet, A. and Dremier, C. (1971) *Experientia 27*, 1003.

13. Hagenmaier, H. and Frank, H. (1972) *Hoppe-Seyler's Z. Physiol. Chem. 353*, 1973.

14. Dorman, L. C. (1969) *Tetrahedron Lett. 28*, 2319.

15. Corley, L., Sachs, D. H. and Anfinsen, C. B. (1972) *Biochem. Biophys. Res. Comm. 47*, 1353.

16. Rony, P. R. (1972) *Biotechnol. Bioeng. Symp. No. 3*, 401.

17. Gordon, A. J. and Ford, R. A. (1972) The Chemist's Companion, Wiley-Intersic. Pub., N. Y.

18. Danckwerts, P. V. (1953) *Chem. Engng. Sci.* 2, 1.

19. Bigelow, C. C. (1967) *J. Theoret. Biol.* 16, 187.

OPTIMIZATION OF ACTIVITIES OF IMMOBILIZED LYSOZYME, α-CHYMOTRYPSIN, AND LIPASE

Rathin Datta and David F. Ollis

Department of Chemical Engineering
Princeton University
Princeton, New Jersey 08540

INTRODUCTION

Two areas of recent development are enzyme immobilization on solid matrices (1) and chemical modification of proteins by soluble reagents (2). Surprisingly, in development of optimum immobilization recipes, there appear to have been no attempts to correlate results between these clearly related areas of enzyme catalysis. As the specific activity (rate per enzyme) of soluble modified enzymes is easily determined by comparison with that of immobilized enzymes, it may be expected that the existence of such correlations would prove a useful screening device for new immobilization recipes. Prior to presenting this study, several pertinent papers are summarized.

Recently Barker and coworkers (4) bound α- and β-amylases on Enzacryls by different coupling methods; they found that α-amylase immobilized through an acid azide coupling method gave the highest specific activity, i.e., activity per unit weight of bound protein. With α-amylase the maximum in specific activity occurred with the catalyst which had the least bound enzyme per gram of carrier. However, β-amylase showed an opposite trend. Moreover, these experiments were not performed with controls which allow distinction between activity changes caused by different chemical coupling methods or better enzyme dispersion on the carrier surfaces.

Martinsson and Mosbach (5) have optimized the binding of pullulanase to an acrylic copolymer by a water soluble carbodiimide: 1-cyclohexyl-3-(2 morpholinoethyl)-carbodiimide metho-p-toluene sulfonate (CMDI). Binding of the protein in presence of the substrate-pullulan gave a five-fold increase in specific activity of the bound enzyme versus the specific activity resulting when the substrate was absent during immobilization. A broad maxima of the bound enzyme activity occurred at an intermediate concentration of CMDI during binding.

Zabriskie, Ollis and Burger (6) observed that covalently bound wheat germ agglutinin on polyacrylamide retained its activity towards binding cell walls; the maximum binding capacity occurred when the concentration of the protein coupling groups on the polyacrylamide bead surface was minimal.

Datta, Armiger and Ollis (7) have shown that immobilized lysosyme on polyacrylamide possessed highest specific activity towards Micrococcus Lysodeikticus cell walls when the density of the binding groups on the bead surface was lowest. They used N-acetyl-α-D-glucosamine (NAG) to protect the active site during immobilization in order to eliminate direct covalent binding at the active site. A model of enzyme de-activation with excessive binding to the polymer matrix was postulated to rationalize such behavior.

None of the above investigations examined changes in specific activity or structure of the soluble proteins when treated by soluble analogs of the same chemical groups used to bind the protein to the support surface. This paper compares the activity of three different enzymes -- lysozyme, α-chymotrypsin and hog pancreatic lipase in their immobilized and their chemically modified soluble forms. Correlations are obtained between changes of specific activity and conformation of the modified enzymes and their activities in the immobilized forms.

MATERIALS AND METHODS

Materials used included the polyacrylamide support Bio-gel P-2 (Bio-Rad Laboratories), lysozyme (hen egg white), α-chymotrypsin (bovine pancreas), lipase (hog pancreas), and dried cells of Micrococcus Lysodeikticus (Worthington Biochemical Corp.). All other chemicals were reagent grade.

Protein, cell, and BTEE concentrations were determined by a Cary-14 spectrophotometer. Specific activities of both soluble and immobilized lipase were measured in a Sargent recording pH stat. The circular dichroism (CD) spectra of protein solutions were obtained with a Cary 60 spectropolarimeter and a CD attachment.

EXPERIMENTAL PROCEDURE

Preparation and Activity of Lysozyme-Polyacrylamide

The procedure of Datta, Armiger and Ollis (7) was used. The kinetic data for lysozyme-polyacrylamide presented in this paper are from the above reference.

Diazotization of Soluble Lysozyme: Lysozyme was diazotized by reaction with diazobenzene sulfonic acid (8). 0.01 mole of sulfanilic acid was dissolved in 250 ml of cold 0.4N HCl solution and 0.05 mole of sodium nitrite was added. The reaction was allowed to proceed for 5 min under constant stirring at 0°C. At the end of 5 min, 1 ml, 0.5 ml, 0.1 ml, 0.05 ml, and 0.01 ml of the reaction solution was added to different beakers, each containing 10 ml of 1 mg/ml lysozyme solution in pH 9 borate buffer. A few drops of 2N NaOH was added to the beakers containing 1 ml and 0.5 ml of the diazobenzene sulfonic acid in order to neutralize the excess acid. As a control, 1.0 ml of cold 0.4N HCl containing 0.2×10^{-3}M of sodium nitrite was added to 10 mg of lysozyme in pH 9 buffer. All beakers were stored overnight at 5°C.

Activity of Soluble Lysozyme: The activity of the diazotized soluble lysozyme and the control was determined from the rate of lysis of one batch of Micrococcus Lysodeikticus in a pH 7 phosphate buffer. The rate was assumed proportional to the decrease in absorbance at 450 mµ of a 0.3 mg/ml suspension of the cells in the buffer. To 2.9 ml of the cell suspension, 0.1 ml of the enzyme solution was added. One unit is taken as the decrease in absorbancy of 0.001 per minute.

Preparation and Activity Measurement
of α-Chymotrypsin-Polyacrylamide

The acyl azide intermediate method of Inman and Dintzis (9) was used to immobilize α-chymotrypsin to Bio-gel P-2. The polyacrylamide beads were reacted with hydrazine hydrate (6M) for 2 hr, 4 hr, 6 hr, 15 hr, 18 hr, and 24 hr to give different surface concentrations of hydrazide groups on the bead surfaces as determined by titration (9). After formation of the acyl azide derivative by reacting with 0.1M sodium nitrite in 0.25N HCl solution at 0°C for 2 min, the acyl azide derivative was washed with cold 0.001N HCl-0.1M $CaCl_2$ solution and suspended in 1 mg/ml solution of α-chymotrypsin in 0.001N HCl-0.1 ml $CaCl_2$. Then 1M NaOH was added to bring the pH to ∼9.0; this value being maintained during reaction by continuous addition of 1M NaOH. The coupling reaction was continued for 1 hr.

The beads were washed by 0.001N HCl-0.1M $CaCl_2$ solution and then by 2N NaCl to remove the non-covalently attached protein. The amount of α-chymotrypsin in the wash is measured by its absorbance at 280 mμ in the spectrophotometer. From the initial amount of enzyme and the final amount in the wash, the attachment of enzyme is calculated.

The activity of immobilized α-chymotrypsin was measured by its action on benzoyl-L-tyrosine ethyl ester (BTEE) (Hummel (10)). 25 ml of 0.00107M of BTEE in 50% (w/w) methanol was mixed with 25 ml of pH 7.9 tris buffer. 1.6 ml of 0.001N HCl-1M $CaCl_2$ was also added. After stirring, 1.6 ml of the solution was pipetted into the spectrophotometer reference cell. The solution in the reaction beaker was then connected through a fitted glass filter to a continuous flow cell in the sample compartment through a Cole-Parmer masterflex pump. The solution was pumped through the cell at a constant flow rate and the instrument was balanced to zero reading at 256 mμ. Now 0.5 g of wet α-chymotrypsin-polyacrylamide beads were added to the reaction beaker and stirred vigorously. Care was taken that the fitted glass filter immersed in the beaker was not clogged by the beads. O.D. readings were taken every 30 secs for 10-15 minutes. After reaction, the enzyme beads in the beaker were filtered and dried to give the exact weight of beads added. The specific activity was calculated by the formula:

$$\frac{\text{units}}{\text{mg attached enzyme}} = \frac{\Delta A_{256}/\min \times 100 \times 50}{964 \times \text{mg attached enzyme}}$$

since 964 equals the molar extinction coefficient for N-Benzoyl-L-tyrosine.

Acylation and Activity of Soluble α-Chymotrypsin

α-Chymotrypsin was acylated by the method of Oppenheimer et al. (11). Five beakers of 100 mg α-chymotrypsin in pH ∼8.5 borate buffer containing 0.05M $CaCl_2$ were cooled to 4°C. Then 0.25, 0.1, 0.05 and 0.01 ml respectively of acetic anhydride was added to four different beakers; pH was maintained between 6.7 and 7.0 by addition of 2N NaOH. After reacting for one hour, excess acetic anhydride (if any) was neutralized by the alkali and the reacted enzyme solutions were dialized overnight against 0.01M sodium borate solution at 4°C. The fifth beaker served as control. The solutions were later diluted to 1 litre in pH 9 borate buffer, and the activities of 0.1 mg/ml protein solutions were measured by the reaction on BTEE according to the method of Hummel (10).

Preparation and Activity Measurement of Immobilized Lipase

Lipase was immobilized on polyacrylamide by Lieberman (12) using the diazonium intermediate method of Inman and Dintzis (9).

The immobilized enzyme was assayed at pH 8 using sonically dispersed tributyrin substrate (see below) with NaOH as a pH-stat titrant.

Diazotization and Activity Measurement of Soluble Lipase

The diazotization of soluble lipase followed the same method as used in the diazotization of soluble lysozyme.

The activity of the soluble diazotized lipase was also assayed by its reaction on tributyrin prepared as follows. A mixture of 1 vol tributyrin is diluted by 50 vols of deionized water and emulsified by sonication for 1 hour. A 5 ml volume of the emulsion is then diluted by 5 ml of deionized water, and to it are added 2 ml of 0.075M calcium chloride solution; 1 ml of 15 mg/ml sodium taurocholate solution and 2 ml of 1M NaCl solution. The 15 ml reactant solution is adjusted to pH 8 and 1 ml solution of the soluble diazotized enzyme of appropriate concentration is added. Rate measurements were again taken with a recording pH stat.

Circular Dichroism Spectra of the Soluble Enzymes

The circular dichroism (CD) spectra of diazotized lysozyme and lipase, acylated α-chymotrypsin and the various native and control samples were measured from 250 to 210 mμ. All the solutions were in pH 9 borate buffer. Most measurements used a 0 sensitivity of 0.1° full scale and a time constant of 3. Below 220 mμ, some samples were too absorbent and scanning was discontinued whenever the instrument noise level became too high. The concentration of the protein in most of the samples was 0.1 mg/ml, and the cell CD path length was 1 cm. The ellipticities are expressed in molecular ellipticities:

$$[\theta]^{25} = \frac{\theta M}{10 \ell c} \left(\frac{\text{degrees cm}^2}{\text{decimole}}\right)$$

where

θ = measured ellipticity in degrees
M = molecular weight of protein
ℓ = path length in cm
c = protein concentration in g/ml

RESULTS

Figure 1 (7) summarizes the kinetic data for the lysis of Micrococcus Lysodeikticus cell walls by lysozyme immobilized on polyacrylamide. The aminoethylation times are marked on each curve. The variation of initial slopes indicates that the specific activity (conversion/mg enzyme/time) of the immobilized lysozyme decreases as the number of protein binding groups on the bead surface increases. There is a small increase in enzyme attachment with increased aminoethylation (Table 1).

The activity of the soluble diazotized lysozyme is found to vary with increasing degree of diazotization as shown in Figure 2. Lysozyme retains its original activity until the ratio of the diazobenzene-sulfonic acid to that of the enzyme reaches one; it then decreases logarithmically with further ratio increases. From the data in Table 2 the enzyme has lost virtually all its activity when the molar ratio (diazo reagent/enzyme) reaches about 30. The specific activity of the enzyme in the control beakers which have received 1.0 ml of the nitrous acid solution did not change from that of the native lysozyme.

OPTIMIZATION OF IMMOBILIZED ENZYME ACTIVITY

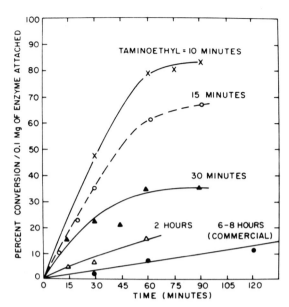

Figure 1: Specific conversion of Micrococcus Lysodeikticus (% conversion/0.1 mg bound enzyme) versus time for various aminoethylation times of: 10 minutes (x), 30 minutes (▲), 2 hours (Δ), and 6-8 hours (·). The dotted line (aminoethylation time, 15 minutes) is for a different batch of enzyme and cell walls.

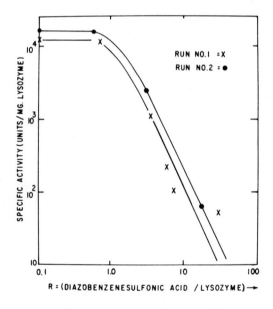

Figure 2: Activity of diazotized lysozyme vs. molar ratio of diazobenzenesulfonic acid to lysozyme.

TABLE 1

CHARACTERIZATION OF POLYACRYLAMIDE-LYSOZYME CATALYSTS

Type of Beads	Duration of Amino-ethylation (min)	H+ Bound to Aminoethylated Beads $\frac{\text{m-moles}}{\text{g-dry wt.}}$	Enzyme Bound $\frac{\text{mg}}{\text{g-dry wt.}}$	m-moles H+/m-moles enzyme bound
Bio-gel P-2 (50-100 mesh)	10	0.0284	2.85	143
"	30	0.1347	4.9	397
"	120	0.808	5.8	2000
Bio-gel P-2* aminoethyl (50-100 mesh)	6-8 hrs	2.0†	10.8	2670

* Commercial Product
† Information obtained from manufacturer's catalogue (Bio-Rad Laboratories)

TABLE 2

LYSOZYME DEACTIVATION DATA

Run 1 Activity of Control = 12,000 units/mg		Run 2 Activity of Control = 16,000 units/mg	
Ratio $\frac{DBS}{lysozyme}$	Activity ($\frac{units}{mg}$)	Ratio $\frac{DBS}{lysozyme}$	Activity ($\frac{units}{mg}$)
0.71	12,000	0.576	16,000
3.55	1,050		
5.75	220	2.88	2,500
7.1	105		
28.8	57	17.3	65
57.6	0	57.6	0

The CD spectra of diazotized lysozyme from 250 to 220 mµ with different degrees of diazotization are plotted in Figure 3. The CD spectra in 6M GuHCl represents the denatured lysozyme spectra. A distinct continuous change in the CD spectra is evident as the degree of diazotization is increased.

The characteristics of the polyacylamide-α-chymotrypsin catalyst are presented in Table 3. The immobilized enzyme has an activity of 1-10% of that of the soluble form. Figure 4 shows the specific activity of the immobilized α-chymotrypsin with increasing concentration of the surface hydrazide groups on the polyacrylamide beads. A maximal specific activity occurs when the hydrazide group concentration is near 2.5 m-moles/g dry wt. This result is different from those of polyacrylamide-lysozyme where the specific activity of the immobilized lysozyme decreased monotonically with increased number of surface attachment groups. Figure 5 is a plot of the specific rate of BTEE hydrolysis by the immobilized α-chymotrypsin versus the enzyme loading (mg enzyme/g dry wt beads). A distinct decrease in specific activity occurs as the enzyme loading increases. Table 3 reveals that chymotrypsin loading reaches a minima with increased derivatization of the beads. The variation in specific chymotrypsin activity with soluble reagent coupling is shown in Figure 6.

TABLE 3
CHARACTERIZATION OF POLYACRYLAMIDE-α-CHYMOTRYPSIN CATALYSTS

Hydridization Time (hr)	Conc. of groups m-moles/gm dry wt.	Enzyme Attached mg/g dry wt.	Units/mg Enzyme Attached
Run #1:			
2	0.112	2.0	0.326
4	0.845	2.18	0.447
6	1.31	1.64	0.79
18	2.49	2.38	1.0
24	4.46	2.84	0.63
Run #2:			
4	0.985	2.5	0.57
6	1.51	1.05	1.08
15	2.46	0.55	3.75
18	3.0	0.57	2.90
24	3.79	2.7	0.60

OPTIMIZATION OF IMMOBILIZED ENZYME ACTIVITY

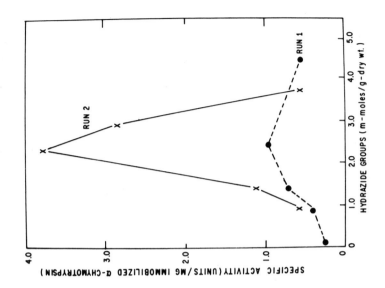

Figure 4: Specific activity of immobilized α-chymotrypsin vs. concentration of surface hydrazide groups on Bio-gel P-2 beads.

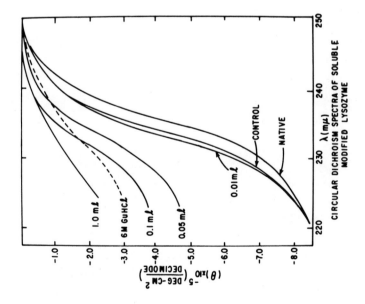

Figure 3: CD spectra of modified lysozyme from 250 mμ to 220 mμ.

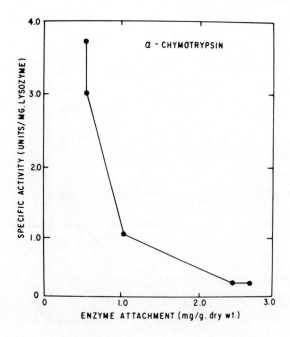

Figure 5: Specific activity of immobilized α-chymotrypsin vs. enzyme loading (mg enzyme per gram of dry beads).

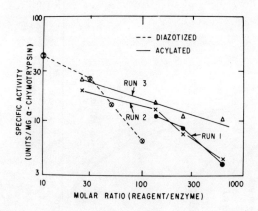

Figure 6: Activity of modified α-chymotrypsin vs. molar ratio of reagent to enzyme. Acylation: solid lines, Run 1 ⊙ , Run 2 x, Run 3 Δ. Diazotization: dashed line ⊗ .

TABLE 4

DEACTIVATION DATA FOR α-CHYMOTRYPSIN

ml. of Acetic Anhydride Added to 100 mg of Enzyme	Ratio Anhydride / Enzyme	Activity (units/mg enzyme)		
		Run #1	Run #2	Run #3
0.25	635	3.74	4.05	10.0
0.1	250	8.7	8.1	13.8
0.05	125	10.9	12.8	14.9
0.01	25	--	19.3	25.0
0	0	38.0	40.0	58.0

Table 4 summarizes the activity of the soluble acylated α-chymotrypsin. There is a gradual monotonic decrease in the α-chymotrypsin activity with increasing acylation for all the three experimental runs.

The CD spectra of acylated α-chymotrypsin is presented in Figure 7. While the degree of acylation caused no significant change in the CD spectra for all the acylated samples, all of them showed a lesser magnitude of ellipticity than the control (which was not acylated) at both the local minima of 229 mμ and the local maxima at 228.5 mμ. This behavior is probably due to some small conformational change that occurs when the enzyme is first acylated. The spectra of α-chymotrypsin in 8M urea shows a very significant decrease in the ellipticity magnitude.

The characterization of immobilized lipase on polyacrylamide is due to Lieberman (12). From the values in Table 5, an increase in the surface coupling groups on the beads leads to a decrease in the specific activity of the bound lipase. This behavior is very similar to that observed with immobilized lysozyme.

Diazotized soluble lipase also shows a large loss in activity as the degree of diazotization is increased (Table 6). The nitrite in the control samples decreased the activity of the soluble enzyme by 60%, and the decrease continued further with increased diazotization (Figure 8).

The CD spectra from 250 to 210 mμ of the lipase samples with different degrees of diazotization do not show any difference in ellipticity. Even the denatured lipase in 8M urea exhibits no significant difference in ellip-

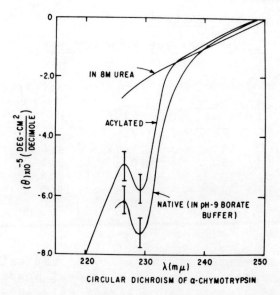

Figure 7: CD spectra from 250 mμ to 220 mμ of modified α-chymotrypsin.

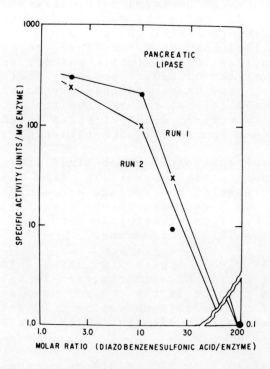

Figure 8: Activity of diazotized lipase vs. molar ratio of diazobenzenesulfonic acid to lipase.

TABLE 5

CHARACTERIZATION OF POLYACRYLAMIDE-LIPASE CATALYSTS

Aminoethylation Time (min)	H^+ Bound to Aminoethylated Beads m-moles / g dry wt.	Enzyme Bound mg / g dry wt.	Avg. Activity units/mg protein
10	0.04	3.38	1.04
30	0.058	2.05	0.41
60	0.16	2.54	0.041

TABLE 6

LIPASE DEACTIVATION DATA

Amt. of Diazo Solution Added (ml)	Ratio Diazo/ Enzyme	Lipase Activity (units/mg) Run #1	Run #2
0.01	2.0	302	230
0.05	10.0	213	78
0.1	20.0	9.58	29.4
1.0	200.0	0	0
Control with 0.1 ml HNO_2 Solution	0	200	200
Control with 1 ml HNO_2 Solution	0	190	190
Native	0	494	500

ticity from the native sample. The lipase used in these experiments was quite impure and the impurities may have masked the actual CD spectra.

DISCUSSION

The activity changes of the soluble enzymes are due to the covalent modification of the amino acid residues. The reagents for the covalent modification used in these experiments were nitrous acid, diazobenzenesulfonic acid and acetic anhydride. A summary of the principal reactions possible with the residues is presented in Table 7.

For the nitrous acid reaction, N_2O_3 is the principal reactant and leads to the formation of an unstable diazonium intermediate which in turn leads to the elimination of the amine group. The α amino groups in proteins are much more susceptible to deamination by nitrite than the ε-amino groups (13, 14). The diazonium reagent reacts primarily with the tyrosine, histidine and lysine residues at the pH of the experiments reported here (2). Acetic anhydride reacts with the α and ε amino groups very readily. It also reacts with the tyrosyl groups but the product is very unstable at pH ~7 where the reaction was carried out (15).

The circular dichroism spectra in the UV region for the diazotized and acylated enzymes are expected to reflect any gross conformational changes in the modified enzymes' structures. The CD spectra of α-helices between 185 to 245 mµ show 3 Cotton bands. These are for the n_1-π^- transition at 206 mµ, and the perpendicularly polarized π°-π^- transition at 190 mµ (16). The CD spectra of β-sheets in silk fibroin show a CD minimum at 218 mµ and a maxima at 197 mµ (17). The random coil conformations, however, show no optical activity between 250 to 220 mµ. Below 220 mµ it shows a slight positive mean-residue ellipticity which becomes negative around 210 mµ and continues to decrease for shorter wavelengths. Figure 9 is a compilation of some pertinent CD spectra of α-helix random coil and β-sheet structures. The CD spectra from 250 to 220 mµ for lysozyme in 6M GuHCl and α-chymotrypsin in 8M urea (Figures 3 and 7 respectively) indicate that the effect of denaturation leads to a drastic change in the conformational structure.

The CD spectra of the various diazotized lysozyme, lipase and acylated α-chymotrypsin suggest the following general conclusions. The lysozyme sample diazotized by

TABLE 7

CHEMICAL REACTIONS FOR ENZYME MODIFICATION

1. **Reactions with nitrous acid:**

$$2HONO \rightleftharpoons N_2O_3 + H_2O$$

$$Protein-NH_2 + \overset{O}{\underset{}{N}}-ONO \xrightarrow{-NO_2^{\ominus}} Protein-\overset{H}{\underset{H}{N^{\oplus}}}-N=O$$

$$\rightarrow Protein-N-N=\overset{\oplus}{O}H \xrightarrow{-H^{\oplus}} Protein-N\equiv N-OH^{\ominus}$$

$$\xrightarrow{-OH^{\ominus}} Protein-N\equiv N \overset{H^{\oplus}}{\Bigg\{} \begin{array}{l} \xrightarrow[-N_2]{H_2O} Protein-OH + H^{\oplus} \\ \xrightarrow[-N_2]{X^-} Protein-X \\ \xrightarrow{-N_2} Protein(u) + H^{\oplus} \end{array}$$

2. **Reaction with diazobenzenesulfonic acid:**

   a. $Protein-\langle O \rangle-O^{\ominus} + 2N_2^{\oplus}-\langle O \rangle-SO_3H \xrightarrow[0°C]{pH \sim 9}$

   $Protein-\langle O \rangle \begin{array}{c} N=N-\langle O \rangle-SO_3H \\ O^- \\ N=N-\langle O \rangle-SO_3H \end{array}$

   b. $Protein-\begin{array}{c} N \\ \Vert \\ N \\ H \end{array} + 2N_2^{\oplus}-\langle O \rangle-SO_3H \xrightarrow[0°C]{pH \sim 9}$

   $Protein-\begin{array}{c} N \\ \Vert N \\ N \\ H \end{array} \begin{array}{c} =N-\langle O \rangle-SO_3 \\ \\ N=N-\langle O \rangle-SO_3 \end{array}$

   c. $Protein-NH_2 + N_2^{\oplus}-\langle O \rangle-SO_3H \longrightarrow$

   $Protein-N \begin{array}{c} N=N-\langle O \rangle-SO_3H \\ N=N-\langle O \rangle-SO_3H \end{array}$

3. **Acetic Anhydride**

$$Protein-NH_2 + CH_3-\overset{O}{\underset{}{C}}-O-\overset{O}{\underset{}{C}}-CH_3 \xrightarrow{pH>7}$$
$$Protein-NH-\overset{}{\underset{O}{C}}-CH_3 + CH_3COO^{\ominus} + H^{\oplus}$$

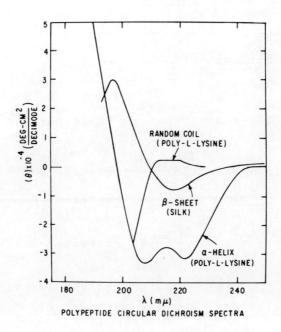

Figure 9: CD spectra of α-helical, β-sheet and random coil polypeptide conformations. Adapted from Holzworth and Doty (16) and Yang (17).

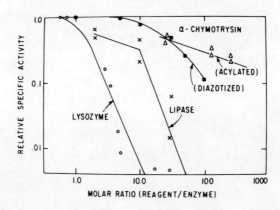

Figure 10: Relative activity of modified lysozyme, lipase and α-chymotrypsin vs. molar ratio of reagent to enzyme.

1 ml of diazobenzene-sulfonic acid solution has not only lost its enzymatic activity but also its ordered conformational structure. The other lysozyme samples with decreasing amounts of diazobenzenesulfonic acid solution show a steady trend toward the native structure along with increase in enzymatic activity. This behavior of lysozyme upon reaction with additional diazonium groups constitutes direct proof for our earlier postulate (8) that lysozyme structure is unfavorably affected upon excessive diazo-intermediate coupling, and specific activity therefore diminishes.

Lipase appears to show no change of CD spectra even though there is a significant loss in enzymatic activity with increased diazotization. This result may be due to the impurities in the lipase sample, which could mask changes of enzyme CD spectra.

α-chymotrypsin is acylated in a molar excess of the reagent which far exceeds that used in diazonium reactions for lysozyme and lipase. Only a small drop in activity occurs accompanied by a very small change in the conformational structure.

The relative loss of specific activity of the soluble enzymes with different degrees of reaction are plotted on a log-log graph in Figure 10 (21). The maximum slopes of the curves and the points of intersection (R_o) with the abcissa at relative activity equal to one are tabulated in Table 8. The slopes for diazotized lysozyme and lipase are much higher in magnitude than that for acylated α-chymotrypsin. The diazotized α-chymotrypsin also shows a much lower decrease in activity with increased diazotization compared to lysozyme and lipase (Figures 6, 10).

The behavior of the three enzymes when immobilized on polyacrylamide by the diazo-coupling and acyl azide methods is seen to be similar to that in the soluble state. Figure 1 and Table 5 show that the activity of the bound enzyme decreases monotonically for both lysozyme and lipase as the number of available enzyme binding groups increases. Figure 10 and Table 8 show that the decrease in activity is substantial for both lysozyme and lipase when the diazonium salt to enzyme ratio is greater than 1 for lysozyme and over 10 for lipase. There is a distinct correlation between the behavior of diazotized lysozyme and lipase and their immobilized forms.

This relation is most evident from Figure 11 (21), which is a plot of relative specific activity of the im-

TABLE 8

ENZYME DEACTIVATION PARAMETERS

Enzyme Name	Intercept (Ro)	Slope (n)
Lysozyme	0.62	-2.26
Lipase	0.4*	-2.58†
α-Chymotrypsin	2.8	-0.35

* Intercept of flat line segment.
† Slope of steep line segment.

mobilized enzymes or the molar ratio of active coupling groups in the beads to the enzyme attached. The slopes of the curves are -0.92 for lysozyme and -1.68 for lipase. On comparing these values to the slopes in the case of soluble enzymes (Table 8: -2.26 for lysozyme and -2.58 for lipase), a clear correlation is obvious. The number of active groups in the beads in Figure 11 is the total number, whereas the enzymes attach only to the superfi-

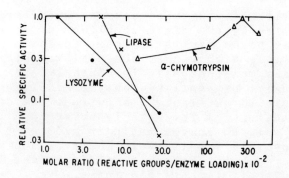

Figure 11: Relative specific activity of the immobilized enzymes vs. the molar ratio of reactive groups on beads to enzyme attached.

cial bead surface. This method of plotting will not change the slopes of the curves from those obtained if only external surface group concentrations were used. A lower slope for the case of each immobilized enzyme than that for the same soluble enzyme is consistent with effects of stearic hindrances during the enzyme binding.

α-chymotrypsin on the other hand shows a different behavior on acylation. The slope of the relative activity and reagent to enzyme ratio in Figure 10 is only -0.35 compared to -2.26 for lysozyme and -2.58 for lipase. Also the acylated α-chymotrypsin shows a substantial activity when the molar ratio (reagent/enzyme) is over 100, whereas all the activity is lost below a (reagent/enzyme) ratio of 30 for lysozyme and about 100 for lipase. The behavior of immobilized α-chymotrypsin shows a maximum near a concentration of 2.5 m-moles/g dry wt for the hydrazide groups. Since the soluble acylated α-chymotrypsin shows no maxima in activity (in the molar range of the experiments), the maxima in the immobilized case may be rationalized as follows. There is not a significant decrease in activity as the number of bonds between the enzyme and the polymer matrix increases as observed from Figure 6. The α-chymotrypsin may bind non-specifically and inactively on a non-derivatized bead surface. Then progressive derivatization may reduce the proportion of non-specific inactive α-chymotrypsin. After exhibiting maximum specific activity the further drop is due to the decrease in activity of the specifically bound α-chymotrypsin with increasing acylation. Figure 7 shows that a lower dispersion of the enzyme on the bead surface causes an increase in the specific activity.

Examination of the values of R_o for the three enzymes in Table 8 shows that R_o is of the order of magnitude of one. This suggests that these enzymes if immobilized by one side group onto the bead surface will probably retain most or all of their activity. For lysozyme and lipase there is also a section of the curve (Figure 10) for low molar ratios where the magnitude of the slope is small. Immobilization in this region of ratios will probably also lead to a fair retention of specific activity.

The object of this study was to examine whether the information easily obtained from modification of soluble enzymes could be of assistance in optimizing enzyme immobilization recipes. From the comparisons summarised in Figures 10 and 11, a clear correlation between the behavior of the modified enzymes in their soluble state and their immobilized forms emerge. This result may be of general utility to other researchers contemplating new immobilization recipes.

ACKNOWLEDGMENT

This work has been supported by the Schultz Foundation and the National Science Foundation (NSF-GI-35996).

REFERENCES

1. Goldman, R., Goldstein, L., and Katchalski, E., in "Biochemical Aspects of Reactions on Solid Supports", G.R. Stark, Ed., p. 1, Academic Press, New York, 1971.

2. Means, G.E., and Feeney, R.E., "Chemical Modification of Proteins", Holden-Day, Inc., 1971.

3. Silman, I.H., and Katchalski, E., Ann. Rev. of Biochem. 35, 873 (1966).

4. Barker, S.A., Somers, P.J., Epton, R., and Melaren, J.V., Carbohydrate Res. 14, 287 (1970).

5. Martensson, K., and Mosbach, K., Biotech. and Bioengg. 14, 715 (1972).

6. Zabriskie, D., Ollis, D., and Burger, M.M., Biotech. and Bioengg. (in press).

7. Datta, R., Armiger, W., and Ollis, D.F., Biotech. and Bioengg. (in press).

8. Haurowitz, F., "Immunochemistry and the Biosynthesis of Antibodies", p. 18, Interscience, 1968.

9. Inman, J.K., and Dintzis, H.M., Biochemistry 8, 10, 4074 (1969).

10. Hummel, B.C.W., Can. J. Biochem. Physiol. 37, 1393 (1959).

11. Oppenheimer, H.L., Labouesse, B., and Hess, G.P., J. Biol. Chem. 241, 2720 (1966).

12. Lieberman, R., and Ollis, D.F. (to be published).

13. Belenkii, B.G., and Drestova, V.A., Biokhimiya 30, 878 (1965).

14. Maeda, H., and Ishida, N., Biochem. Biophys. Acta 147, 597 (1967).

15. Fraenkel-Conrat, H., Methods Enzymol. 4, 247 (1959).

16. Holzwarth, G., and Doty, P., JACS 87, 218 (1965).

17. Yang, J.T., "Conformation of Biopolymers", Vol. 1, G.N. Ramachandran, ed., p. 157, Academic Press, 1967.

18. Kagan, H.M., and Vallee, B.L., Biochemistry 8, 11 (1969).

19. Saxena, V.P., and Wetlaufer, D.B., Biochemistry 9, 25, 5015 (1970).

20. Pecheré, J., Dixon, G.H., Maybury, R.H., and Neurath, H., J. Biol. Chem. 233, 1364 (1958).

21. Ollis, D.F., Datta, R., and Cox, E.C. (submitted to Science).

CHEMICAL MODIFICATION OF MUSHROOM TYROSINASE FOR STABILIZATION TO REACTION INACTIVATION

David Letts* and Theodore Chase, Jr.

Department of Biochemistry and Microbiology

Cook College, Rutgers - The State University

New Brunswick, New Jersey 08903

INTRODUCTION

The utilization of chemical modification to evoke or favor a particular response while avoiding an undesirable characteristic of the protein is nothing new. Formaldehyde has been used to modify bacterial toxins, rendering them incapable of eliciting a toxic response, but still able to produce an immunological response when injected into an animal. The ancient process of tanning has been improved by the use of glutaraldehyde, a cross-linking agent. Insolubilization of enzymes for use in column or filtration techniques for the destruction or alteration of substances has received much recent attention. Hence, immobilization and subsequent modification of an enzyme, or immobilization of the modified enzyme, would seem to be a reasonable procedure.

We were seeking to use mushroom tyrosinase (o-diphenol:O_2 oxidoreductase, EC 1.10.3.1) as an immobilized enzyme for the synthesis of L-Dopa from L-tyrosine (Fig. 1). L-Dopa is presently a drug of choice in the treatment of Parkinson's disease, being a natural intermediate in the formation of dopamine, which is apparently deficient in certain critical areas of the brain in this condition. The DL mixture obtained by chemical synthesis gives rise to undesirable side effects, and a process using readily available optically pure L-tyrosine as starting material should have considerable value.

---

*Present address: General Foods, Inc., Tarrytown, N.Y.

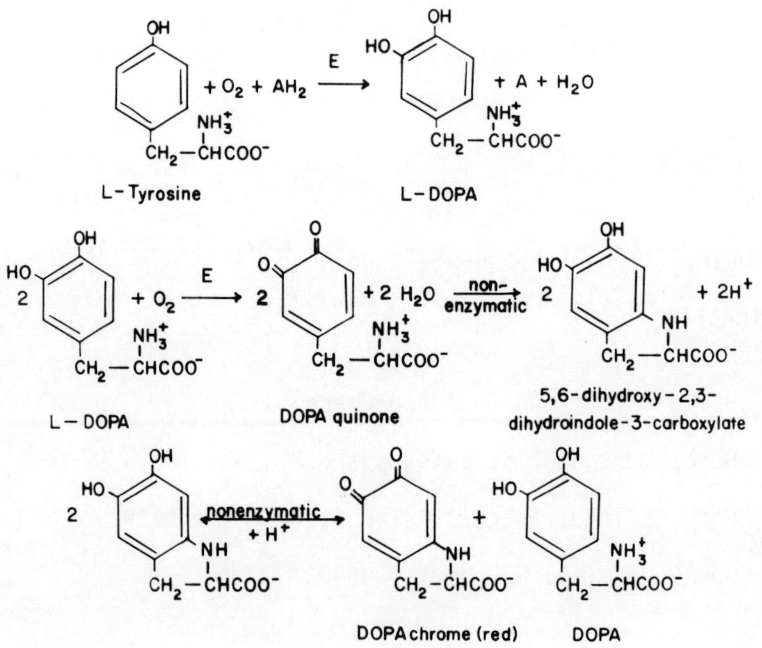

Figure 1. Action of tyrosinase on L-tyrosine and L-Dopa, and non-enzymatic reactions of dopaquinone.

However, as the figure shows, the enzyme catalyzes not only hydroxylation of monophenols, the desirable reaction for our purpose, but also the oxidation of the dihydroxy product to an o-quinone, which can undergo further non-enzymatic reactions. The second reaction must be minimized if a good yield of the diphenol is to be obtained. Secondly, the reaction is seriously handicapped by gradual inactivation of the enzyme as the reaction proceeds. This "reaction inactivation" phenomenon is due to covalent attachment of the substrate or product to the enzyme (Wood and Ingraham, 1965). These authors suggested that a likely explanation for the effect is nucleophilic attack of lysine amino groups of the protein on the quinone product, yielding a covalent adduct which blocks the active site. We have, therefore, sought to stabilize the enzyme to reaction inactivation by chemical modification of lysine residues, in order to use the modified enzyme, immobilized on a collagen membrane, for synthesis of L-Dopa.

We have used the immobilized enzyme in a plug-flow reactor, both because it is a likely configuration for large-scale Dopa production and because immediate removal of the product Dopa from the enzyme should decrease further oxidation of it and, hypothetically, minimize reaction inactivation.

MATERIALS AND METHODS

Mushroom tyrosinase was purchased from Worthington Biochemical Corporation. In experiments reported here we have used this preparation without further purification, although we and others have been able to purify the enzyme about twenty-fold further by ammonium sulfate precipitation and column chromatography.

Since we were interested in Dopa production, not catechol oxidation, we used the method of Arnow (1939) for determination of Dopa formed from tyrosine. The enzyme was assayed in 0.5 mM L-tyrosine, 0.1 M K phosphate buffer pH 7.0, usually containing 1 mM Na ascorbate. To 1 ml assay mixture after 5 minutes incubation with enzyme, or to 1 ml samples from batch or flow assay of the immobilized enzyme, was added 0.25 ml 2N HCl, then 1 ml 10% $NaNO_2$ - 10% Na_2MoO_4; the solutions were mixed thoroughly with a Vortex mixer (thorough aeration is necessary in presence of ascorbate), and 0.25 ml 4N NaOH was added to develop a red color. The optical density was read at 495 nm in a Spectronic 20 colorimeter; an absorbance of 0.425 corresponded to 0.1 µmole Dopa. A fresh solution of mushroom tyrosinase (770 Worthington units/mg) formed 0.255 µmoles Dopa per min per mg under these conditions. Assay solutions were always prepared freshly with solid Na ascorbate.

Immobilization of mushroom tyrosinase on collagen was accomplished according to the method of Wang and Vieth (1973). A collagen membrane, 1242 cm^2 and 0.36 mm thick, was swollen in 0.05 M Tris Cl buffer, pH 8, and soaked in 44 ml of mushroom tyrosinase solution, 10 mg/ml, at 4°C for 24 hr. During this time the enzyme penetrates into the swollen membrane. The film was then dried and stored at 5°C. Suitably sized strips were cut off, modified by the particular chemical, and introduced into the reactor module or assayed batchwise. Non-collagen protein immobilized on the membrane was determined by tryptophan content (Eskamani *et al.* 1973).

Modification of the collagen-immobilized enzyme was carried out with glutaraldehyde, ethyl acetimidate, and dimethyl adipimidate. The glutaraldehyde modification was accomplished by soaking a 12 x 3 cm piece of the immobilized enzyme film in 25 ml of 2.3%

glutaraldehyde solution, buffered at pH 6 with 0.1 M K phosphate, for 15 hr. For modification with ethyl acetimidate and dimethyl adipimidate, twenty equivalents of the hydrochloride of the compound per equivalent of lysine [based on a lysine content of 4.5%, as determined for the purified enzyme (Jolley *et al.* 1969)], plus Na_2CO_3 sufficient to neutralize the hydrochloride, were added to 10 ml of 0.1 M K phosphate or $NaHCO_3$ buffer, pH 8.5, containing collagen immobilized enzyme film (generally 12 x 3 cm) and placed on a shaker for 1.5 hr at 25°C. The film was then removed from the reaction solution and washed in 0.1 M K phosphate buffer, pH 7.

Assay of the immobilized enzyme was carried out in two ways. In the batch assay, a sample of enzyme-bearing membrane was put into assay solution (55 ml, usually 0.5 mM L-tyrosine - 1 mM Na ascorbate in 0.1 M K phosphate buffer pH 7, in a 150 ml beaker). The solution was mixed and aerated by steady bubbling with air, and at appropriate times 1 ml samples were withdrawn for determination of Dopa content. In the flow assay, the reactor module (Fig. 2) was fed a substrate solution (concentrations as above), usually pumped through at a rate of 3 ml/min. At appropriate times a 1 ml sample of the effluent was taken for determination of the Dopa content.

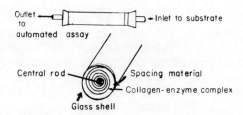

Figure 2. Flow reactor for assay of membrane-bound tyrosinase.

RESULTS

We expected that the greatest problem in Dopa synthesis by immobilized tyrosinase would be "reaction inactivation", the gradual loss of activity in presence of the substrate; this effect defeated the similar efforts of Wykes *et al.* (1971). The effect can be seen in Fig. 3: the enzymatic activity is reasonably stable when stored in buffer alone, but decreases sharply over a period of a few hours in the presence of tyrosine. Since the most probable mechanism for "reaction inactivation" is reaction of nucleophilic groups on the enzyme with the quinone product, we first tried adding sodium ascorbate to the reaction mixture, to reduce any quinone present back to Dopa, thus maximizing Dopa production and hopefully minimizing "reaction inactivation."

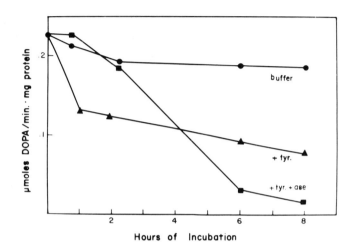

Figure 3. Loss of activity of mushroom tyrosinase when stored in presence of tyrosine. 0.5 mg commercial enzyme was incubated at 23°C in 1 ml 0.1 M K phosphate buffer pH 7, unsupplemented (●), supplemented with 0.5 mM L-tyrosine (▲), or with 0.5 mM L-tyrosine - 1 mM Na ascorbate (■). At the indicated times samples were withdrawn and assayed for enzymatic activity (Dopa formation).

However, although initial activity of the enzyme was stimulated, "reaction inactivation" was not inhibited significantly.

A second approach to the problem is immobilization of the enzyme and use of a flow reactor, from which the Dopa product is continually removed, in the hope that keeping the Dopa concentration low will minimize "reaction inactivation." We were able to immobilize tyrosinase on a collagen membrane by the method of Wang and Vieth, although the initial activity of the immobilized enzyme was very low compared to the free enzyme, particularly in the batch assay (Table 1). This may in part be due to preferential adsorption of other proteins - 42% of the protein was removed from

TABLE 1

Initial rate of action of free and immobilized tyrosinase

Free enzyme	15.3 μmoles/mg · hr
Immobilized enzyme, batch assay	0.085 μmoles/mg · hr
Immobilized enzyme, flow reactor	0.47 μmoles/mg · hr

Activity and non-collagen protein were determined as described in the text.

the tyrosinase solution in which the membrane was incubated, but only 25% of the activity. We have not as yet been trying to maximize enzyme bound. Familiar problems such as reduced accessibility of substrate to the enzyme, change of K_m and overlapping of membrane are probably also important, especially in the batch assay.

When immobilized enzyme was incubated in a beaker of tyrosine solution, Dopa production ceased after four to six hours, even in presence of ascorbate or at low pH (Fig. 4). This was not simply attainment of equilibrium between synthesis and oxidation of Dopa, since if the membrane were removed at this time, washed, and placed in a fresh tyrosine solution, it failed to form further Dopa. Presence of borate, which might be expected to complex with the vicinal hydroxyls of Dopa, did cause some persistence of activity. Use of the immobilized enzyme in a flow reactor increased the initial activity five-fold, but activity was virtually absent after 8.5 hr (Fig. 5). This loss was not primarily due to leaching of protein from the membrane, since over half the non-collagen protein remained on the membrane at this time, and twenty-four hours of washing the membrane with phosphate buffer did not reduce the immobilized activity appreciably.

Therefore, we sought to modify the enzyme chemically, in the hope of preventing the hypothesized attack of nucleophilic groups upon the reactive product. We selected ethyl acetimidate as a modifying reagent which would make lysine ε-amino groups less reactive, both by raising their pKa and by steric hindrance (Fig. 6). Maintaining a positive charge on lysine side chains is desirable for maximum stability of the enzyme; in this connection, it may be noted that modification with succinic anhydride, which replaces the positive charge with a negative charge, yielded an active enzyme, but one which speedily lost activity even on storage in buffer, probably due to dissociation, as noted by Jolley *et al.* (1969).

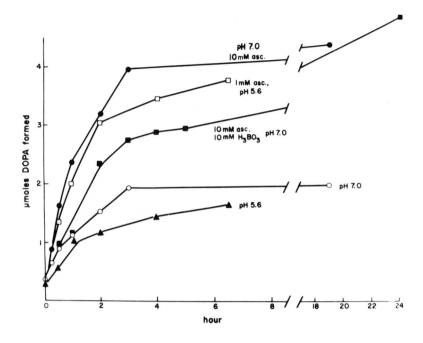

Figure 4. Synthesis of Dopa by immobilized enzyme in batch assay under various conditions. Samples (42 cm^2) of immobilized enzyme were assayed by the batch procedure, using 0.9 mM L-tyrosine in 0.09 M K phosphate buffer pH 7.1, with additions as follows: o, no addition; △, Na succinate, 0.09 M pH 5.6, in place of K phosphate pH 7.1; ●, Na ascorbate, 10 mM, added; ▲, Na ascorbate, 1 mM, and Na succinate, 0.09 M pH 5.6, in place of K phosphate pH 7.1; ■ , Na ascorbate, 10 mM, and H_3BO_3, 10 mM, added.

It should be noted that collagen membranes are peculiarly appropriate substrates for immobilization of a chemically modified enzyme, since the immobilization is stated to depend on hydrogen bonds, salt linkages and van der Waals interactions (Wang and Vieth, 1973), rather than upon covalent bonds involving lysine amino groups, as is the case with most other methods of immobilization. Chemical modifications of lysine ε-amino groups therefore may prevent immobilization by covalent bonding to these groups, but not immobilization on collagen.

Ethyl acetimidate modification actually increased activity (Fig. 7). Unfortunately, we were not able to get satisfactory results for decrease of free amino groups; ethyl acetimidate itself reacts with trinitrobenzenesulfonic acid, and samples passed through

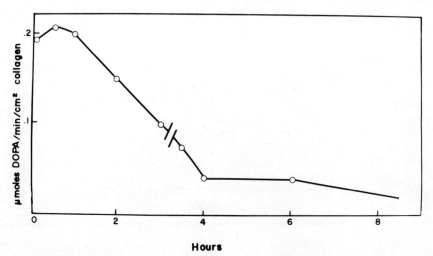

Figure 5. Synthesis of Dopa by immobilized enzyme in flow reactor. Enzyme membrane (60 cm) was placed in the flow reactor (Fig. 2), and L-tyrosine solution (0.5 mM, 0.1 M K phosphate pH 7, 1 mM in Na ascorbate) was pumped through at a rate of 4 ml/min, reduced to 3 ml/min after 3 hr. Samples (1 ml) of the effluent were taken from time to time for determination of Dopa content. At the break between 8 and 8.5 hr, 0.1 M K phosphate buffer pH 7, without tyrosine, was pumped through for 24 hr.

$$RC\overset{NH_2^+}{\underset{OR'}{\diagup}} + H_2NR'' \longrightarrow RC\overset{NH_2^+}{\underset{\|}{-}}NHR'' + R'OH$$

Figure 6. Reaction of imido esters with amines.

Sephadex columns showed no consistent decrease in reactivity with that reagent.

The modified enzyme showed definitely increased stability in storage in presence of tyrosine (Fig. 8). In another experiment, 60% of the initial activity persisted after storage for three days at 5°C in presence of tyrosine.

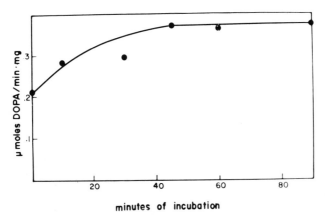

Figure 7. Increase of activity during reaction with ethyl acetimidate. To free enzyme (100 mg, in 5 ml 0.1 M dimethylaminoethanol Cl buffer, pH 8.5) was added, with stirring, ethyl acetimidate HCl (76.1 mg) and Na_2CO_3 (69.3 mg). Samples (5 μl) of the incubation mixture were withdrawn at the indicated times and added to tubes containing 1 ml of the assay mixture. Dopa content of these tubes was determined after 8 min.

We then tried modification of the immobilized enzyme with ethyl acetimidate, with dimethyl adipimidate as a similar reagent which could also introduce covalent cross-links to the collagen, and with glutaraldehyde as another cross-linking reagent. Unfortunately, modification of the immobilized enzyme with the imidate reagents did not result in increased stability (Fig. 9). Again, loss of activity does not seem to be due primarily to loss of protein, since over half the non-collagen protein remained immobilized when all activity had disappeared (Table 2).

TABLE 2

Immobilized non-collagen protein remaining after chemical modification and use in flow reactor.

Membrane	Protein mg/cm^2 membrane
Untreated, unused membrane	1.0
Untreated membrane after 9 hrs of tyrosine flow, 24 hrs wash with buffer	0.55
Membrane treated with ethyl acetimidate, 5 hrs tyrosine flow	0.88
Membrane treated with dimethyl adipimidate, 6 hrs tyrosine flow	0.62

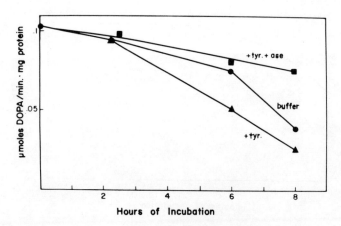

Figure 8. Loss of activity of acetimidate-modified tyrosinase stored in presence of tyrosine. Free enzyme modified with ethyl acetimidate (Fig. 7) and separated from that reagent by gel filtration was incubated (0.3 mg in 1 ml) in 0.1 M phosphate buffer pH 7.0, unsupplemented (●), supplemented with 0.5 mM L-tyrosine (▲) or with 0.5 mM L-tyrosine plus 1 mM Na ascorbate (■). At the indicated times samples were withdrawn and assayed for enzymatic activity.

Glutaraldehyde modification did result in a considerable increase of stability, though initial activity was somewhat diminished. The total amount of Dopa synthesized over 15 hours with the glutaraldehyde-treated membrane increased 50% over that synthesized by the untreated membrane, from 1.35 μmoles Dopa per mg immobilized non-collagen protein to 2.03; this is to be compared with synthesis of 9 μmoles per mg free enzyme in a 24 hr incubation. In addition, attempts to repeat the glutaraldehyde modification have resulted in a membrane with little initial activity, possibly due to displacement of non-collagen protein into solution during the modification reaction, since some tyrosinase activity was found in the glutaraldehyde incubation solution after removal of the membrane.

We intend to try two further approaches: (1) immobilization of enzyme previously modified with ethyl acetimidate; (2) modification of enzyme tyrosine residues as well as lysine, possibly by methylation. The justification for the latter approach lies in the

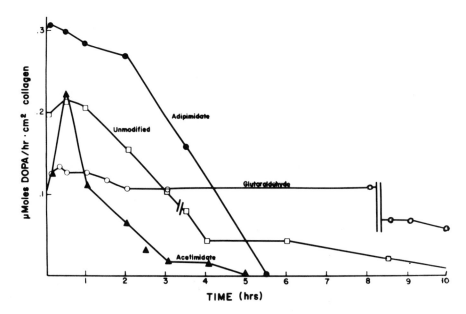

Figure 9. Synthesis of Dopa by chemically modified enzyme in flow reactor. Enzyme membranes modified with ethyl acetimidate, dimethyl adipimidate or glutaraldehyde were placed in the flow reactor, and L-tyrosine solution (0.5 mM, 1 mM in Na ascorbate, 0.1 M K phosphate buffer pH 7) was pumped through at a rate of 3 ml/min. 1 ml samples of the effluent were taken at the indicated times for determination of Dopa content. Results with unmodified enzyme (Fig. 5) are included for comparison. At the breaks shown with unmodified and glutaraldehyde-modified enzyme, 0.1 M K phosphate buffer, without tyrosine, was pumped through (3 ml/min) for 24 hours.

observation of Jolley *et al.* that iodinated tyrosinase was more subject to reaction inactivation than the unmodified enzyme. This is a modification which would lower the pKa of the tyrosine hydroxyls of the enzyme; a modification which would prevent their ionization or nucleophilic reaction might make the enzyme less subject to reaction inactivation.

In summary, we have demonstrated Dopa synthesis by tyrosinase immobilized on collagen membrane, but have not solved the reaction inactivation problem, though some of the results are encouraging.

REFERENCES

Arnow, L.E. 1937. J. Biol. Chem. 118:531.
Eskamani, A., Chase, T., Jr., Freudenberger, J., and Gilbert, S.G. 1973. Analytical Biochemistry, in press.
Jolley, R.L., Jr., Robb, D.A., and Mason, H.S. 1969. J. Biol. Chem. 244:1593.
Wang, S.S., and Vieth, W.R. 1973. Biotechnology and Bioengineering 15:93.
Wood, B.J.B., and Ingraham, L.L. 1965. Nature 205:291.
Wykes, J.R., Dunnill, P., and Lilly, M.D. 1971. Nature New Biology 230:167.

CHAIN REFOLDING AND SUBUNIT INTERACTIONS IN ENZYME MOLECULES COVALENTLY BOUND TO A SOLID MATRIX*

H. Robert Horton and Harold E. Swaisgood

Departments of Biochemistry and Food Science

North Carolina State University, Raleigh, N. C. 27607

The biological and chemical properties of functionally active proteins are dependent upon their three-dimensional molecular structures. Insight into the role of the gene-determined primary sequence of a protein molecule in providing a thermodynamically stable tertiary structure with biological activity was obtained through the classical experiments of White, Anfinsen, and their colleagues on the renaturation of reductively denatured ribonuclease A (1-3). In these and subsequent investigations, it has been demonstrated that the acquisition of functional tertiary structures in relatively simple, single-chained protein molecules can occur spontaneously as a result of the inherent thermodynamics of the system; *i.e.*, interactions among the linear array of amino acid residues and the molecular environment.

However, various difficulties have arisen in analogous experiments with protein molecules consisting of more than one polypeptide chain, such as α-chymotrypsin (three disulfide-linked chains) and lactate dehydrogenase (four dissociable polypeptide subunits). For example, air-reoxidation of denatured, fully reduced chymotrypsinogen A in solution results in a maximum recovery of 1.4% of catalytically active chymotrypsin upon subsequent treatment with trypsin (4). Low recoveries in this and other systems (5,6) appeared to stem from aggregation of denatured polypeptide chains.

In order to examine more fully the regeneration of biologically functional conformation from completely denatured proteins under

---

*This investigation has been supported by National Science Foundation grant GI-39208. Paper 4206 of the Journal Series of the N. C. State University Agricultural Experiment Station, Raleigh, N. C.

conditions which would minimize interactions leading to non-specific aggregation, we have utilized two glass-bound enzyme systems: rabbit muscle lactate dehydrogenase (EC 1.1.1.27) and bovine chymotrypsinogen A (EC 3.4.4.5).

IMMOBILIZED LACTATE DEHYDROGENASE

Rabbit muscle lactate dehydrogenase (composed chiefly of tetramers of the M-type subunit) was chosen as a representative multichained enzyme for examination of the effects of surface immobilization on enzymatic activity and chain refolding. Differences between the apparent Michaelis constants of enzyme immobilized through amide bonds *via* its carboxyl groups ("aminopropyl-glass" enzyme) and those of a similarly modified soluble enzyme (prepared by substituting glycine methyl ester for matrix-bound amino groups) apparently result from the effects of the surface environment and not from the chemical modification *per se* (see Table I). Whereas the small increase in apparent Michaelis constant for NADH may be most easily attributed to restrictive effects on intra-pore diffusion, the sizeable decrease in apparent K_m for pyruvate cannot be explained by such a phenomenon. Such a decrease in value could be the result of a matrix-induced conformational change in the enzyme to increase its affinity for pyruvate, or could result from partition of pyruvate into the matrix to provide an effective concentration greater than that in the bulk solution. The latter possibility would have to involve some form of specific interaction rather than general electrostatic attraction, since glass-enzyme preparations of opposite

Table I. Apparent Michaelis Constants of Native, Modified, and Glass-Bound Lactate Dehydrogenase[a]

Preparation	K_{NADH} (μM)	$K_{pyruvate}$ (μM)
Native	7.8	185
Modified (Gly-OMe)	6.4	154
Succinamidopropyl-glass	39	5.1
Aminopropyl-glass	55	3.6

[a] All measurements were made in 0.1 M phosphate buffer at pH 7.0 and 25°C.

surface charge (succinamidopropyl-glass and aminopropyl-glass) exhibit similarly lowered K_m values for pyruvate (Table I), and since NADH (which, like pyruvate, is negatively charged) shows elevated rather than lowered K_m values.

To measure the number of active sites and to further examine the possibility of matrix-induced conformational change in the immobilized enzyme, the binding of ^{14}C-NAD$^+$ was measured in equilibrium experiments. Evaluation of the data, presented as a Scatchard plot in Figure 1, revealed that all of the theoretical number of binding sites were active (4.27 ± 0.53 sites/enzyme molecule). Even more striking was the value of 29 µM obtained for the dissociation constant, which is thirty-fold less than that of 910 µM which has been reported for native enzyme in solution (7).

Two criteria were examined to assess functional chain refolding: regain of enzymatic activity and ability to bind native subunits from solution. Dissociation and unfolding of the glass-immobilized dehydrogenase was accomplished in 7 M Gdn·HCl.[1] Determination of the ratio of protein (by amino acid analyses) which remained bound to the Gdn·HCl-washed matrix to that which was removed in the denatur-

---

[1] Abbreviations used are: Gdn·HCl, guanidinium chloride; Bz-TyrOEt, N-benzoyl-L-tyrosine ethyl ester; Z-Tyr-ONp, N-benzyloxycarbonyl-L-tyrosine p-nitrophenyl ester; DTNB, 5,5'-dithio-bis(2-nitrobenzoic acid); EDC, 1-ethyl-3-(dimethylaminopropyl)-carbodiimide.

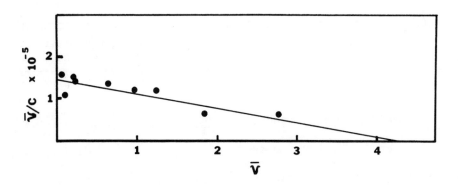

Figure 1. Scatchard Plot of ^{14}C-NAD$^+$ Binding Data. The line was computed by a program designed to give the best least squares fit of the data to a hyperbolic equation for one class of binding sites.

ing solution revealed that three-fourths of the enzyme was dissociable from the glass (see Table II). The ratio, 0.33, represents the minimum theoretical ratio which is obtainable provided only one subunit of each immobilized tetramer had been covalently bound to the matrix. It can also be seen from the data in Table II that a fraction of the non-covalently linked subunits had dissociated from the matrix-bound subunits following two months' storage of the immobilized enzyme in phosphate buffer; that is, the ratio of protein remaining bound to that dissociated in Gdn·HCl was increased from 0.33 to 0.74. Previous studies of rabbit muscle lactate dehydrogenase in solution have shown that storage under similar conditions leads to dissociation of the tetrameric form into dimers (8).

Following removal of the denaturant, equilibration of the glass-bound subunits with substrate solution results in a return of 10-20% of the original enzymatic activity, as shown in Table III. If the regained activity were actually the result of 10% of the tetramers remaining undissociated and covalently attached to the matrix following Gdn·HCl washing, then the ratio of Table II should have been 0.48 rather than 0.33. Such a value is outside the range of experimental error; thus, these results suggest that the individual matrix-bound subunits are capable of exhibiting lactate dehydrogenase activity.

Table II. Ratio of the Amount of Enzyme Remaining on the Glass to That Removed by Washing with 7.0 M Guanidinium Chloride

Preparation	Ratio[a]
Fresh succinamidopropyl-glass-enzyme	0.33 ± 0.02
Succinamidopropyl-glass-enzyme aged two months[b]	0.74 ± 0.04

[a] Ratio is expressed as protein remaining bound/protein in Gdn·HCl wash; the amounts of protein were determined by amino acid analysis of glycine; error is based on a 3% error in amino acid analysis.

[b] Aged preparations were stored in pH 7.0 phosphate (0.1 M) at 4°C.

Table III. Relative Activities of Immobilized Lactate Dehydrogenase Before and After Washing with 7 M Guanidinium Chloride, and Following Reconstitution[a]

Preparation	Stirred Tank Reactor[b] (% Original Activity)			Column Reactor[c] (% Original Activity)		
	Before	After	Reconstituted[d]	Before	After	Reconstituted[d]
Aminopropyl-glass	100	17	108	100	11	114
Succinamidopropyl-glass (2 prepns)	---	--	---	100	10	100
				100	13	89

[a] All activity measurements were made in 0.03 M phosphate buffer, pH 7.0, with pyruvate and NADH as substrates.

[b] Data are expressed as percent of the original activity in ΔA_{340}/min.

[c] Data are expressed as percent of the original activity in ΔA_{340}, which is the difference in absorbance of substrate solution passed through a column of beads with and without bound enzyme.

[d] Reconstitution was accomplished by incubating the Gdn·HCl-washed immobilized enzyme with an aged solution of lactate dehydrogenase (0.1 mg/ml).

Incubation of the refolded matrix-bound subunits with a dilute solution of native enzyme, followed by washings as employed in preparing freshly immobilized enzyme (9), results in complete return of the initial level of enzymatic activity (Table III, "Reconstituted"). Thus, it can be concluded that the matrix-bound, Gdn·HCl-denatured subunits are capable of refolding to produce catalytic activity, individually, and to interact with subunits in solution to produce a completely functional immobilized tetramer.

IMMOBILIZED CHYMOTRYPSINOGEN

Regeneration of tertiary structure in unfolded, glass-bound protein chains was investigated in greater detail using chymotrypsinogen A. As previously noted, attempts to generate biological activity by refolding reductively denatured proteases or zymogens in solution have met with little success. By analogy to the presumed folding of nascent polypeptide chains attached to ribosomes *in vivo*, we postulated that refolding of a denatured chymotrypsinogen chain which was covalently bound to an insoluble matrix should meet with greater success, provided the microenvironment of the surface did not adversely affect the structural interactions of the polypeptide chain (10).

Accordingly, chymotrypsinogen A was covalently bound to EDC-activated succinyl groups of succinamidopropyl-glass beads, and the beads were washed free of residual non-covalently bound protein. A quantity of the succinamidopropyl-glass-bound chymotrypsinogen was reductively denatured by treating with 10% (v/v) β-mercaptoethanol in 8 M urea, 0.05 M in Tris-Cl, pH 8.6, under nitrogen for 18 hr. An alternate, more efficient procedure involved similar treatment for 4-5 hr with 3-4 mM dithiothreitol (in 10- to 20-fold molar excess over the potentially available sulfhydryl groups of the protein) in place of β-mercaptoethanol. Completeness of reductive denaturation in each case was confirmed by washing a portion of the beads free of denaturant and reducing agent with 0.1 M acetic acid (under nitrogen), and then titrating the sulfhydryl groups of the glass-bound protein with DTNB (11). Protein content was determined by amino acid analysis following acid hydrolysis.

The reductively denatured glass-bound zymogen preparation was washed free of denaturant and reducing agent and allowed to reoxidize in 0.05 M Tris-Cl, pH 8.6. Following reoxidation, the glass-immobilized protein was treated with a solution of trypsin at pH 7.8 and then assayed for chymotryptic activity using a variety of substrates. Kinetic data, such as are presented in Figure 2, were analyzed by least squares fitting to the integrated form of the Michaelis-Menten equation (10):

$$C_p = K_m \cdot \ln(1 - f) + k_o E_o V_o / Q$$

where C_p is the product concentration; K_m, the apparent Michaelis constant; f, the fraction of substrate converted to product; k_o, the overall catalytic coefficient of the enzyme; E_o, the total enzyme concentration (determined by amino acid analyses of hydrolyzed bead samples); V_o, the void volume of the column of immobilized enzyme; and Q, the flow rate. Table IV presents the kinetic constants obtained for the "reoxidized" preparation and for trypsin-activated immobilized chymotrypsinogen which had never been exposed to reductant ("native" preparation). Comparison of the bimolecular rate constants (k_o/K_m) of the reoxidized preparation to those of the native preparation revealed a 76-80% recovery of esterolytic activity towards Z-Tyr-ONp, and a 53% recovery of activity towards Bz-TyrOEt. The air-reoxidized preparation's activity towards casein was 30% of that of nonreduced matrix-bound enzyme.

In contrast to these findings concerning the efficacy of refolding of denatured, matrix-bound chymotrypsinogen chains, similar attempts to achieve refolding of reductively denatured matrix-bound

Table IV. Characteristics of Native and Reduced-Reoxidized Succinamidopropyl-Glass-Chymotrypsinogen Activated with Trypsin

Matrix-bound Enzyme	Substrate	$K_m \times 10^5$ (M)	k_o (sec^{-1})	k_o/K_m (% of native)	Specific[a] Activity
Native	Z-Tyr-ONp	1.3	0.22	100	
Reoxidized 1	"	1.1	0.14	76	
Reoxidized 2	"	1.0	0.15	80	
Native	Bz-TyrOEt	12.8	11.0	100	
Reoxidized	"	6.8	3.1	53	
Native	Casein				0.10
Reoxidized	"				0.03

[a]Values for casein specific activities are given as trichloroacetic acid-soluble ΔA_{280}/min/mg glass-bound protein.

chymotrypsin failed to provide any measurable enzymatic activity, thus implying that the entire zymogen molecule is needed for generation of biologically functional protein conformation.

Two methods have been developed to permit further characterization of the generation of tertiary structure in chymotrypsinogen: incorporation of a thioester linkage to bind the protein to the glass beads, which permits subsequent, selective chemical cleavage of the zymogen from the matrix (12); and construction of a solid-phase fluorimetry cell to permit examination of the emission spectra of glass-bound proteins.

In the first procedure, aminopropyl-glass beads were succinylated as previously described (10), and then activated with EDC and treated with thioglycolic acid under an atmosphere of nitrogen. The thioglycolated beads were thoroughly washed with distilled water, then activated with EDC and coupled with chymotrypsinogen as before, to provide the following linkage:

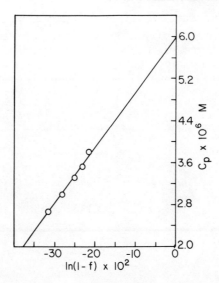

Figure 2. Kinetics of Trypsin-Activated, Air-Reoxidized Succinamidopropyl-Glass-Chymotrypsinogen's Catalysis of Z-Tyr-ONp Hydrolysis at pH 5.0. Based on increase in absorbance at 340 nm due to production of p-nitrophenol.

$$(glass)-CH_2CH_2CH_2-\overset{H}{\underset{|}{N}}-\overset{O}{\underset{\|}{C}}-CH_2CH_2\overset{O}{\underset{\|}{C}}-S-CH_2-\overset{O}{\underset{\|}{C}}-\overset{H}{\underset{|}{N}}-(\text{protein})$$

The zymogen, thus bound, was reductively denatured with dithiothreitol + urea, then allowed to reoxidize in the presence of phosphate buffer, pH 7.0. Solubilization of the immobilized, reoxidized protein was achieved by 40 minutes treatment with a neutral solution of 1.0 N hydroxylamine, which effected cleavage of the thioester bond, thereby releasing thioglycolated zymogen. The protein, thus released, was found to contain 1.1 ± 0.1 moles of SH groups/mole, implying that the conditions employed in attaching chymotrypsinogen to the glass had resulted in the binding of only one amino group per protein molecule. Titration of the released, thioglycolated protein (after trypsin activation) with ^{14}C-diisopropylphosphofluoridate revealed that the specific conformation of the Asp(102)-His(57)-Ser(195) "charge-relay system" had been restored by the air-reoxidation procedure.

Evidence that the overall conformation of thioglycolated protein which had been reductively denatured and reoxidized prior to its release from the matrix was not identical to that of released, thioglycolated chymotrypsinogen which had not been reduced was obtained through comparison of their ultraviolet absorption spectra and fluorescence emission spectra. The reoxidized preparation exhibited a blue shift of *ca.* 10 nm in absorption maximum with respect to that of the nonreduced preparation (280 nm), and a red shift in emission maximum, implying a greater exposure of tryptophanyl residues to a polar environment in the overall conformation of the air-reoxidized samples. These differences may have arisen in part during the hydroxylamine cleavage, however, rather than from refolding of the immobilized polypeptide chain, *per se*, as evidenced by identical solid-phase fluorescence spectra (see Figure 3).

It can thus be concluded that immobilization of chymotrypsinogen has allowed us to circumvent the problem of intermolecular interactions which lead to aggregation in structural regeneration studies. Similar systems may be applicable to other proteins. On the basis of the results obtained with immobilized chymotrypsinogen and lactate dehydrogenase, it appears that the "thermodynamic hypothesis" of Anfinsen and his colleagues (13) may be generally applicable to the generation of functional tertiary and quaternary structures in multichained as well as single-chained protein species.

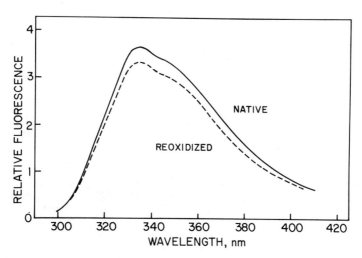

Figure 3. Solid Phase Fluorescence Emission Spectra of Trypsin-Activated Succinamidopropyl-Glass-Chymotrypsinogen ("Native") and Trypsin-Activated Reduced, Reoxidized Succinamidopropyl-Glass-Chymotrypsinogen ("Reoxidized"). Each preparation excited at 280 nm, in 0.1 M acetic acid.

REFERENCES

1. Sela, M., White, F. H., Jr., and Anfinsen, C. B. (1957) *Science* 125, 691.

2. White, F. H., Jr. (1961) *J. Biol. Chem.* 236, 1353.

3. Anfinsen, C. B., and Haber, E. (1961) *J. Biol. Chem.* 236, 1361.

4. Brown, J. C., and Horton, H. R. (1972) *Proc. Soc. Exp. Biol. Med.* 140, 1451.

5. Epstein, C. J., and Anfinsen, C. B. (1962) *J. Biol. Chem.* 237, 3464.

6. Nakagawa, Y., and Perlmann, G. E. (1970) *Arch. Biochem. Biophys.* 140, 464.

7. Fromm, H. J. (1963) *J. Biol. Chem.* 238, 2938.

8. Cho, I. C., and Swaisgood, H. (1973) *Biochemistry* 12, 1572.

9. Cho, I. C., and Swaisgood, H. E. (1972) *Biochim. Biophys. Acta* 258, 675.

10. Brown, J. C., Swaisgood, H. E., and Horton, H. R. (1972) *Biochem. Biophys. Res. Commun.* 48, 1068.

11. Ellman, G. L. (1959) *Arch. Biochem. Biophys.* 82, 70.

12. Brown, J. C., and Horton, H. R. (1973) *Federation Proc.* 32, 496.

13. Anfinsen, C. B. (1973) *Science* 181, 223.

IMMOBILIZATION OF LIPASE

TO CYANOGEN BROMIDE ACTIVATED POLYSACCHARIDE CARRIERS

Paul Melius and Bi-Chong Wang

Department of Chemistry

Auburn University, Auburn, Alabama 36830

INTRODUCTION

Brandenberger attached lipase to the isocyanato derivative of poly-p-aminostyrene in 1956 (1). This was the first time that lipase was immobilized to an insoluble carrier through covalent bond formation. Kitajima also entrapped the lipase in microcapsule (2). The present paper describes the preparation and some properties of lipase bound to polysaccharides following cyanogen bromide activation.

MATERIAL AND METHODS

Porcine pancreatic lipase (EC 3.1.1.3) from this department produced by a procedure first described by Melius (3); Cellulose, Sephadex and Sepharose was from Sigma (USA) and cyanogen bromide was purchased from Fluka (Switzerland).

Coupling of Lipase to Cellulose and Sephadex

Immobilization of lipase to cellulose and Sephadex was done by the cyanogen bromide method (4). The general procedure for coupling of lipase to carrier was as follows:
A weighed amount (300-400 mg) of mercerized cellulose or dry Sephadex was suspended in 16 ml BrCN (25 mg/ ml H_2O) solution. During the 12 min activation the pH was maintained at desired value (pH 9.1-10.0) with 2M NaOH using a pH-Stat. The activated cellulose or Sephadex was then washed with 300 ml of cold 0.1M $NaHCO_3$ to make sure it was free from unreacted BrCN. Then it was

suspended in 0.5 ml lipase solution in a closed test tube and was rotated end-over-end for 16-24 hr at 4°C. The coupling pH value was controlled by using sodium carbonate/sodium bicarbonate buffer solution (pH 7-10.3). The coupled lipase was then washed thoroughly with 0.1M $NaHCO_3$ (24 hr), 1mM HCl (1 hr), 0.5M NaCl (24 hr) and H_2O (3 hr) at a flow rate of 8 ml/hr in a 30 ml Gooch glass filter.

Coupling of Lipase to Agarose

Agarose gels were washed with distilled water in a glass filter. Excess water was removed under suction for 3 min. Six g of swollen gel (about 150 mg dry agarose) was added to 8 ml H_2O and 6 ml BrCN (25 mg/ ml H_2O) solution. The activation and subsequent washing step are the same as for the cellulose (4). Then 1 g swollen (about 25 mg dry) CNBr activated agarose was suspended in 2 ml of buffer solution in a test tube and lyophilized lipase was added directly to this suspension. It was stirred slowly for 20 hr. After the end of the coupling reaction it was washed with 200 ml cold 0.1M $NaHCO_3$ solution and 300 ml of cold distilled water.

Assay of Enzymatic Activity

The enzyme assay procedure used in this work was described by Fritz (5). The Radiometer Automatic Titrator was used in all the enzyme assays. For the assay of immobilized lipase, the weighed amount of dry powder was used directly except for agarose-enzyme conjugate, which was used in the form of a suspension.

Determination of Amount of Bound Lipase

The amount of lipase covalently bound to the polysaccharides was determined by ninhydrin analysis according to Crook (6) after acid hydrolysis.

RESULTS AND DISCUSSION

The degree of activation is somewhat dependent on the activation pH value. Even though there is no clear relationship between coupling and activation pH value, the coupling yield increases a little bit when the activation pH changes only 0.4 from pH 9.1 to pH 9.5 (Table 1). For the same activation pH value, the coupling yield was reduced about 10% while the amount of enzyme used was doubled. The ability of cyanogen bromide-activated polysaccharides to react with proteins under alkaline conditions is in most cases high.

Table 1

Chemical coupling of lipase to cyanogen bromide-activated polysaccharides

Polymer (activated pH) value	Amount of enzyme used per 25 mg dry polymer mg (coupling pH)	Enzyme bound per g dry conjugate mg/g	Coupling yield based on amount of enzyme added (%)
Cellulose(pH 10)	2.4 (pH 9.7)	76.4	84
Cellulose(pH 10)	4.6 (pH 9.7)	86.1	50
Cellulose(pH 10)	5.4 (pH 9.7)	59.5	30
Cellulose(pH 10)	5.6 (pH 9.0)	79.5	38
Cellulose(pH 10)	14.9 (pH 7.0)	82.1	15
Sephadex(pH 10)	2.1 (pH 9.0)	78.0	100
Sephadex(pH 10)	6.0 (pH 7.0)	166.0	83
Sepharose(pH 9.1)	0.7 (pH 9.7)	25.9	95
Sepharose(pH 9.1)	1.8 (pH 9.7)	54.5	80
Sepharose(pH 9.1)	2.2 (pH 9.7)	58.2	70
Sepharose(pH 9.1)	1.2 (pH 10.3)	43.0	90
Sepharose(pH 9.1)	1.5 (pH 10.3)	47.9	86
Sepharose(pH 9.1)	2.4 (pH 10.3)	65.4	73
Sepharose(pH 9.5)	1.2 (pH 9.7)	40.9	89
Sepharose(pH 9.5)	1.5 (pH 9.7)	46.3	81
Sepharose(pH 9.5)	2.5 (pH 9.7)	72.5	77
Sepharose(pH 9.5)	1.8 (pH 10.3)	55.9	82

For Agarose and Sephadex, the coupling yield can be as high as 95% or more of added enzyme. Compared with Agarose and Sephadex, cellulose needs a more basic environment for its activation. Exposure of the enzyme to a reactive polymer with high density of reactive groups increases the risk of multiple attachment, which might tend to decrease the activity of enzyme (4). As shown in Table 1, the coupling yield decreases very rapidly when the amount of enzyme used increases. The same is true for total activity retained. It seems that the ratio of cellulose to enzyme should be greater than 10 to retain one quarter of its activity. The Agarose to enzyme ratio should be greater than 30 to retain 10% of its activity.

Table 2

Activity of bound lipase toward tributyrin substrate (TB)
or triolein substrate (TO)
(continuation of Table 1)

Polymer (activated pH) value	Activity toward TB or TO		Total activity retained (%)
	pH optimum	Activity ratio (%)	
Cellulose(pH 10)	10.4	26.2 (TO)	22.0
Cellulose(pH 10)	10.4	10.4 (TO)	5.2
Cellulose(pH 10)	10.4	13.3 (TO)	4.0
Cellulose(pH 10)	10.4	1.8 (TO)	0.7
Cellulose(pH 10)	-	- (TO)	-
Sephadex(pH 10)	-	- (TO)	-
Sephadex(pH 10)	-	- (TO)	-
Sepharose(pH 9.1)	9.0	9.9 (TB)	9.4
Sepharose(pH 9.1)	8.0	2.0 (TB)	1.6
Sepharose(pH 9.1)	9.0	3.0 (TB)	2.1
Sepharose(pH 9.1)	9.0	6.1 (TB)	5.5
Sepharose(pH 9.1)	9.0	3.4 (TB)	2.9
Sepharose(pH 9.1)	9.0	2.3 (TB)	1.7
Sepharose(pH 9.5)	9.0	4.8 (TB)	4.3
Sepharose(pH 9.5)	9.0	3.2 (TB)	2.6
Sepharose(pH 9.5)	9.0	2.5 (TB)	1.9
Sepharose(pH 9.5)	9.0	1.8 (TB)	1.5

Because of the steric hindrance of enzyme-carrier conjugate, the activity is strongly dependent on which kind of substrate was used. For the Agarose-carrier conjugate, there is almost no activity at all toward triolein substrate (data not shown in the table). This is because the matrix of conjugate is too crowded to allow the large substrate to reach the active sites of the enzyme.

For cellulose-enzyme conjugate, the pH optimum shift is about 1.4 more basic (from pH 9.0 to pH 10.4), but no shift was observed for the Agarose-enzyme conjugate.

All the data in this paper imply that cyanogen bromide activated polysaccharides probably may not be suitable for lipase immobilization because of steric hindrance and too great of a loss in enzyme activity. This might be solved by introduction of an arm between the carrier and enzyme.

By using 4,4'-Methylenedianiline (MDA) as a spacer, the modified amino-containing Sepharose 2B derivative was used as support for immobilization of lipase. It was found that when using carbodiimide as coupling reagent in DMF, the activity retained is at least three times that of the corresponding preparation without introducing the MDA spacer.

REFERENCES

1. H. Brandenberger, Rev. Ferment. Ind. Aliment., 11, 237, 1956.
2. M. Kitajima, S. Miyano, and A. Kondo, Kogyo Kagaku Zasshi, 72, 493, 1969 (C.A., 70, 118067a, 1969).
3. P. Melius, and W.S. Simmons, Biochim. Biophys. Acta. 105, 600-602, 1965.
4. R. Axen, and S. Ernback, Eur. J. Biochem., 18, 351, 1971.
5. J. Fritz, and P. Melius, Canad. J. Biochem. Physiol., 41, 719 1963.
6. E.M. Crook, K. Brocklehurst, and C.W. Wharton, Methods in Enzymology, 19, 963, 1970.

USE OF IMMOBILIZED ENZYMES FOR SYNTHETIC PURPOSES

David L. Marshall

Battelle, Columbus Laboratories

505 King Avenue, Columbus, Ohio 43201

INTRODUCTION

Although considerable interest has been shown in the application of immobilized enzyme technology, comparatively little attention has been devoted to those enzymes or enzyme systems involved in synthesis of complex substances from simple precursors. For very valid economic reasons, the main industrial interest has been on enzymes which catalyze degradative reactions such as starch or protein hydrolysis.

In our laboratory, we have been investigating the use of immobilized enzyme technology for the synthesis of some form of edible food from metabolic wastes. Our interest is in determining the feasibility of this process for use in future long-term space flights. In this report, we describe our results to date with this approach.

SELECTION OF REACTION PATHWAYS

The selection of pathways for the synthesis of edible food from metabolic wastes can, in theory, include a wide variety of possible starting materials, intermediates and final products. For this study, however, we have limited the selection based on certain conditions associated with the unique problem of prolonged space flight. Some of the conditions which influenced the choice of reactions are listed below.

o Combustion of metabolic wastes will provide CO_2 as the starting chemical

o Carbohydrate rather than protein or fat is the desired final product

o Starch-like polysaccharide is the main type of carbohydrate to be synthesized

o Cofactors and/or non-edible co-products need to be recyclable

o Energy considerations not likely to be as important as total weight of processing equipment

With these and other considerations, two major pathways have so far been proposed. One route is a reconstruction of the photosynthetic carbon assimilation process using enzyme-catalyzed reactions all the way from CO_2 fixation to polysaccharide synthesis. The second route uses non-enzymatic reactions to synthesize a three-carbon compound such as glyceraldehyde or dihydroxyacetone and enzyme-catalyzed reactions for the remaining reactions. The two routes become identical once the triose phosphates are obtained. These reactions are summarized in Figure 1.

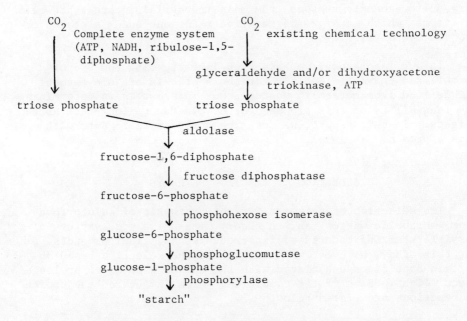

Figure 1. Pathways From CO_2 to Polysaccharide

USE OF IMMOBILIZED ENZYMES FOR SYNTHETIC PURPOSES

The enzyme-catalyzed CO_2 fixation route is considerably more complex, requiring more different steps, a greater number of enzymes and a far greater need for cofactor regeneration. The increased cofactor need is probably the most serious problem. For every mole of glucose equivalent produced, 18 moles of ATP and 12 moles of NADH are required. In addition, the 5-carbon CO_2 recipient, ribulose-1, 5-diphosphate, must be regenerated. This means that 5/6 of the 6-carbon products has to be used for restructuring the CO_2 recipient. This imposes great separation problems to the entire process.

A much simpler pathway uses non-enzyme catalyzed reactions to convert CO_2 up to a level of complexity at which enzyme specificity is needed. In this case, this means three-carbon compounds such as glyceraldehyde, dihydroxyacetone or perhaps glycerol. The required non-enzymatic reactions have not been investigated in this project but are known in the chemical literature. The general pathway involves conversion of CO_2 into formaldehyde, acetaldehyde, acrolein and glyceraldehyde. Glyceraldehyde or dihydroxyacetone would be preferred over glycerol since no further oxidation is needed. This would eliminate the need for NADH in the entire process. ATP still would be required, but only at the level of 2 or 3 moles per mole of glucose equivalent produced, depending on the final choice of reactions.

In addition to the choice of routes leading from CO_2 to hexose phosphate, a choice also exists for conversion of hexose phosphate into polysaccharide. The possibilities are indicated in Figure 2.

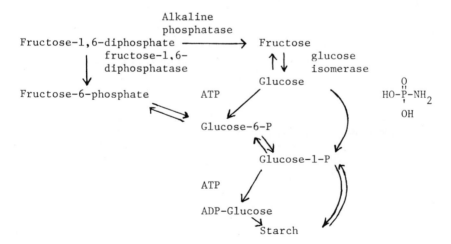

Figure 2. Alternate Routes From Hexose Phosphate to Starch

Our studies have included various portions of each of the major pathways. These reactions, which are summarized below, constitute the bulk of the discussion which follows.

- o Polysaccharide synthesis using immobilized phosphorylase

- o An enzymatic procedure for ATP regeneration

- o Phosphorylation of glucose using ATP and phosphoramidate as phosphoryl donor

- o Two-enzyme system (and two cofactors) for converting 3-phosphoglyceric acid into glyceraldehyde-3-phosphate

- o Three-enzyme system for converting triose phosphate into fructose-6-phosphate

METHODS

Glyceraldehyde-3-phosphate Dehydrogenase
(GAPDH) and Phosphoglycerate Kinase (PGK)

Assay. Activity of GAPDH was followed spectrophotometrically by measuring the decrease in absorbance at 340 nm resulting from the utilization of reduced NADH. The other required substrate, 1,3,-diphosphoglyceric acid, was generated *in situ* by the action of PGK on 3-phosphoglyceric acid and ATP. To a 1 cm light path cuvette were added the following solutions: 1.0 ml 0.15 M imidzaole-HCl, pH 7.0, 0.4 ml 7.5 mM $MgCl_2$, 1.0 ml 3mM ATP, 0.25 ml 0.208 mM NADH, 0.1 ml 0.15 M 3-phosphoglyceric acid (PGA), 0.1 ml 0.015 M dithioerythritol and 1 µg phosphoglycerate kinase. After standing 10 minutes, reaction was started by adding suitably diluted GAPDH.

One unit of activity is defined as the amount of enzyme which will oxidize 1 µmole NADH per minute at room temperature.

The assay procedure for PGK was identical to the GAPDH procedure except for the substitution of excess soluble GAPDH for excess PGK. The time lag used before starting the GAPDH assay was unnecessary in the PGK assay.

For the various immobilized GAPDH derivatives, the assay procedure was modified as follows. Double quantities of each solution used in the soluble GAPDH assay were added to a 15-ml vial mounted outside the spectrophotometer. With a Sigmamotor pump, the assay solution was circulated through a flow-through cuvette. A nylon net filter on the inlet tubing prevented solid particles from entering the spectrophotometric cell.

This method was later replaced by a simpler and more direct assay technique using a Beckman Acta III spectrophotometer and is described later in this section.

Enzyme Source. Many of the initial experiments were performed with crystalline enzymes (Sigma). Later, a crude mixture of the two enzymes was obtained from yeast. A low-temperature dried bottom brewer's yeast (10 g) was mixed with 30 ml 0.1 M $NaHCO_3$ and allowed to autolyze overnight at room temperature. Then, 75 ml cold H_2O was added and the extract centrifuged. To 50 ml extract was added slowly 15.7 g $(NH_4)_2SO_4$. The mixture was centrifuged and the precipitate discarded. An additional 7.0 g $(NH_4)_2SO_4$ was added to the supernatant. After stirring for 30 minutes, the precipitate was collected by centrifugation and saved for subsequent experiments.

Immobilization. A large number of immobilization techniques was tried for either GAPDH or PGK alone or for a mixture of the two. In the following descriptions, only one use of any technique is given in detail.

CNBr-Activated Sepharose[1]

Sepharose 6B was washed well with deionized H_2O using a fritted-glass funnel. A portion of the washed suspension (4 ml) was transferred to a test tube and mixed with 2 ml CNBr solution (50 mg/ml). The pH was immediated adjusted to 11.0 with 4N NaOH. The contents were constantly mixed with a magnetic stirrer while maintaining the pH at 11.0 with periodic addition of 4N NaOH. At the end of 8 minutes, the activated Sepharose was poured into a fritted-glass funnel and washed thoroughly with cold H_2O. After one wash with cold 0.1 M $NaHCO_3$, the Sepharose was resuspended in 4 ml 0.1 M $NaHCO_3$ and 2 ml added to a GAPDH solution (2.5 mg crystalline enzyme from rabbit muscle in 3.0 ml 0.1 M $NaHCO_3$). The mixture was stored in the refrigerator overnight. The Sepharose was washed free of excess GAPDH by a continuous wash procedure. The Sepharose was packed into a small column and washed with a total of 100 ml 0.05 M imidazole-HCl at a flow-rate of approximately 1 ml/min.

Adsorption on Colloidal Silica[2]

A solution of GAPDH (2.5 mg in 3 ml 0.05 M acetate buffer, pH 6.0) was added to 10 mg silica and stirred for 30 minutes. The mixture was centrifuged and the silica pellet resuspended in 3 ml acetate buffer. Then, 10 mg of 1-ethyl-3-(3-dimethylaminopropyl) carbodiimide was added and left to stand overnight at 4°C. The silica was washed several times with acetate buffer.

Adsorption on Ion-Exchangers

Samples (500 mg) of dry DEAE- and CM-Sephadex were swollen by warming in H_2O at 60-80°C for 2 hours. Each type of Sephadex was washed with H_2O and then 0.05 M imidazole buffer, pH 7.0. After resuspending the solids in imidazole buffer, 12 mg of each type was added to solutions of GAPDH (2.5 mg in 2.5 ml imidazole buffer). The product was washed several times with imidazole buffer to remove nonadsorbed enzyme.

Alkylamine Glass and Glutaraldehyde[3]

Porous, alkylamine glass (100) mg was suspended in 3 ml H_2O and treated with 0.1 ml 50 percent glutaraldehyde. After 30 minutes, the glass was washed 8-10 times with H_2O and then once with 0.05 M imidazole buffer, pH 7.0. To the washed glass was added 4.0 ml GAPDH solution. For the next 2 hours, the glass was occasionally mixed on a vortex mixer. The product was washed several times with imidazole buffer and assayed with the flow-through system.

Iodoacetylcellulose[3]

Iodoacetylcellulose was prepared by the exchange reaction of bromoacetylcellulose with iodine. To 1 g bromoacetylcellulose, 6 g sodium iodide and 50 ml 95 percent ethanol were added and the mixture stirred at 30°C for 20 hours. The solid was filtered and washed thoroughly with ethanol, 0.1 M $NaHCO_3$, H_2O and ethanol, and dried.

For the preparation of bromoacetylcellulose[5], 10 g cellulose powder was dispersed in a solution of bromoacetic acid (100 g) dissolved in 30 ml dioxane and the mixture was stirred at 30°C for 16 hours. To the mixture, 75 ml of bromoacetylbromide was added and the stirring continued at 30°C for 7 hours. Then the mixture was poured into a large volume of H_2O, filtered and washed successively with H_2O, 0.1 M $NaHCO_3$ and H_2O.

For GAPDH immobilization, the enzyme was first treated with dithiobis(2-nitro-benzoic acid) (DTNB). The enzyme (~1 mg in 0.05 M phosphate buffer, pH 8.5) was added to 10 mg iodoacetylcellulose along with enough solid ammonium sulfate to give a final 1 M concentration. The mixture was stirred overnight at 4°C and then washed several times with phosphate buffer followed by several washes with 0.05 M imidazole buffer, pH 8.0. The solid was suspended in 1 ml of imidazole buffer and treated with 0.1 ml 0.15 M dithioerythritol. After 30 minutes, the product was washed again and assayed.

$TiCl_4$-Activated Cellulose[6]

One g microcrystalline cellulose was mixed with 5 ml titanic chloride and placed in a desiccator containing NaOH. The desiccator was evacuated and placed in an oven at 45°C. After the drying was complete, a portion of the solid (10 mg) was washed three times with 0.05 imidazole buffer and added to 3 ml imidazole buffer containing 1 mg GAPDH. The 280 nm absorbance of the solution was monitored for 3 hours. At the end of this time, the cellulose was washed well and assayed.

Entrapment in Polyacrylamide Gel[7]

To solution of 1 mg crystalline yeast PGK in 1.0 ml 0.05 M imidazole buffer pH 7.0 was added 0.285 g acrylamide and 0.015 g N,N-methylene bisacrylamide. Nitrogen gas was slowly bubbled into the solution for 5 minutes. Ammonium persulfate (0.01 g) and N,N,N,N-tetramethylethylenediamine (0.016 ml) was added and the solution poured into a stirred mixture of 9.7 ml toluene, 3.7 ml $CHCl_3$ and 0.035 ml sorbitan trioleate. The entire mixture was stirred at 240 rev/min for 30 minutes at 4 C under a stream of N_2 gas. The gel beads were then filtered and washed with 0.1 M $NaHCO_3$, H_2O, and 0.05 M imidazole buffer pH 7.0.

Bio-Gel P-150 Hydrazide[8]

Ten g dry polyacrylamide beads (Bio-Gel P 150) were allowed to swell overnight in 240 ml H_2O. The flask containing the swollen Bio-Gel and another flask containing 60 ml hydrazine was placed in a water bath at 48-50°C. After 30 minutes, the hydrazine was added to the Bio-Gel which was stirred with a magnetic stir bar. At the end of 2 hours, 4 hours, and 7 hours, one-third portions of the mixture was removed and washed well on a funnel with 0.1 M NaCl. Subsequent washings with H_2O were accomplished by sedimentation and decantation.

Portions of each Bio-Gel derivative (300 mg), representing varying degrees of hydrazide substitution, were treated with 45 ml 0.3 N HCl. After cooling to 3°C, 3 ml 1M sodium nitrite was added to each sample. At the end of 10 minutes, the Bio-Gel was washed on a Buchner funnel with cold 0.3 N HCl and then cold H_2O. The gel was transferred to a solution of 8 ml crude PGK/GAPDH mixture in 0.05 M borate buffer, pH 8.5 (350 total units). Stirring was maintained 1 hour at 3°C. Portions of the supernatant were removed for assay. To convert unreacted acid azide groups into amide group, 215 ml 3 M NH_4Cl-1M NH_4OH were added and stirring continued for 1 hour. Portions of each gel sample were packed in a small column and washed continuously with borate buffer and then 0.05 M imidazole buffer, pH 7.0 containing 0.5 mM dithioerythritol. The gel was then assayed using the batch, flow-through technique.

Encapsulation[9]

The $(NH_4)_2SO_4$ precipitate containing both PGK and GAPDH was suspended in 70 percent saturated $(NH_4)_2SO_4$ solution and encapsulated with a cellulose acetate butyrate membrane according to the following procedure:

Polymer: 3.0 g cellulose acetate butyrate, dissolved in 100 ml toluene at stirring speed 75 rpm for 5 minutes. Then 250 ml methylene chloride was added and stirring continued 5 minutes.

Core Material: 30 ml 70 percent saturated $(NH_4)_2SO_4$ containing suspended PGK/GAPDH precipitate (approximately 600 total units) and 0.5 percent bovine serum albumin.

Reaction Vessel: Preparation of microcapsules done in 100 ml beaker surrounded by 4°C ice bath. Stirring accomplished with an overhead, 3-blade stirrer mounted in the center of the beaker.

Step I. Add core material to polymer solution that is already mixing; disperse core solution at 300 rpm for 2 minutes. All subsequent stirring speeds at 300 rpm.

Step II. Add 100 ml toluene (from graduated addition funnel) dropwise over a 30-minute period. Steps III-VII also employ dropwise addition from a graduated container.

Step III. Add 100 ml toluene - 20 minutes.

Step IV. Add 100 ml toluene - 15 minutes.

Step V. Add mixture of 100 ml toluene - 25 ml petroleum ether (high boiling) 25 minutes.

Step VI. Add mixture of 62.5 ml toluene and 62.5 ml petroleum ether - 20 minutes.

Step VII. Add 125 ml petroleum ether - 15 minutes.

Step VIII. Stop stirrer and let capsules settle; decant off liquid until 400 ml remain.

Step IV. Add 250 ml petroleum ether (all at once) and stir for 30 minutes at 300 rpm.

Step X. Stop stirrer and let capsules settle. Decant off liquid until 400 ml remain.

Step XI. Add 250 ml petroleum ether as in Step IX and stir 15 minutes at 300 rpm.

Step XII. Filter capsules on sintered glass funnel.

Step XIII. Permit capsules to air dry for 1 hour then humid air dry for 4 hours.

The capsules from this process contain a concentrated $(NH_4)_2SO_4$ core solution. Gradual removal of the salt was accomplished by slowly adding H_2O (at 4°C) to a stirred 70 percent saturated $(NH_4)_2SO_4$ suspension of capsules. When enough H_2O has been added to lower the $(NH_4)_2SO_4$ concentration to 5 percent saturation, the capsules were filtered and washed with cold H_2O.

Continuous operation of PGK/GAPDH columns. To determine the stability of the various immobilized PGK/GAPDH derivatives, substrate solutions were continuously pumped through columns containing the immobilized enzyme. The substrate solution consisted of 0.05 M imidazole buffer, pH 7.0, 1mM $MgCl_2$, 1 mM dithioerythritol, 0.05 mM EDTA, 0.2 mM ATP, 0.2 mM PGA, and 0.2 mM NADH.

An appropriate amount of each derivative was placed in the column so that the percent conversion could be varied between 50-100 percent by changes in flow-rate between 0.05-0.5 ml/min. The proper amount of immobilized enzyme was as little as 0.1 ml in the case of PGK/GAPDH - Sepharose and as much as 2 ml in the case of the glass derivatives. The column effluent was measured at periodic intervals for 340 nm absorbance. When the activity of the column decreased to approximately one-half of the initial activity, the experiment was either terminated or, in some cases, the column was treated overnight with a solution of soluble PGK/GAPDH.

Triose phosphate isomerase (TPI), Fructose
diphosphate aldolase (aldolase) and
Fructose-1,6-diphosphatase (FDPase)

<u>Assay procedures for individual enzymes</u>. TPI was assayed by using glycerophosphate dehydrogenase to convert dihydroxyacetone phosphate into α-glycerophosphate together with the conversion of NADH to NAD. The following solutions were added to a 1 cm cuvette: 1.5 ml 0.1 M glycine buffer, pH 9.4, 0.25 ml 2.5 mM NADH, 0.02 ml 12.5 mM D,L-glyceraldehyde phosphate, 0.05 ml glycerophosphate dehydrogenase (5 units), and 1.17 ml H_2O. Reaction was started by adding 0.01 ml TPI solution (crystalline yeast enzyme from Sigma).

One unit of activity is defined as the amount of enzyme which will convert 1 μmole glyceraldehyde-3-phosphate to dihydroxyacetone phosphate per minute at room temperature.

The assay procedure for aldolase uses fructose diphosphate as the substrate and measures the rate of cleavage into triose phosphate. The appearance of triose phosphate is detected spectrophotometrically using the combination of triose phosphate isomerase, α-glycerophosphate dehydrogenase and NADH.

To a 1 cm cuvette were added the following solutions: 1.5 ml 0.1 M glycine buffer, pH 9.4, 0.5 ml 0.06 M fructose-1,6-diphosphate, 0.25 ml 2.5 mM NADH, 0.01 ml TPI (2 units), 0.1 ml α-glycerophosphate dehydrogenase (10 units), and 0.63 ml H_2O. Reaction was started by the addition of 0.01 ml aldolase solution (crystalline rabbit muscle enzyme from Sigma).

One unit of activity is defined as the cleavage of 1 μmole fructose diphosphate/minute at room temperature. Since one fructose diphosphate produces two triose phosphates, the measured rate of NADH oxidation is divided by two.

FDPase catalyzes the irreversible hydrolysis of fructose-1,6-diphosphate to fructose-6-phosphate and inorganic phosphate. Assay of the enzyme can be based on measurement of the appearance of either fructose-6-phosphate or inorganic phosphate. For systems involving only FDPase, the assay involves the measurement of fructose-6-phosphate using the coupled enzyme pair, phosphohexose isomerase and glucose-6-phosphate dehydrogenase, and the appearance of NADPH.

To a 1-cm cuvette were added the following solutions: 1.5 ml 0.1 M glycine buffer, pH 9.4, 0.3 ml 10 mM $MnCl_2$, 0.15 ml 10 mM NADP, 0.2 ml phosphohexose isomerase (2.5 units), 0.2 ml glucose-6-phosphate dehydrogenase (5.0 units), 0.05 ml 10 mM fructose-1,

6-diphosphate, and 0.95 ml H_2O. Reaction was started with the addition of 0.01 ml FDPase solution (rabbit liver preparation from Sigma).

For immobilized derivatives of the individual three enzymes, the assay was performed in the same way as for the soluble enzyme. This was made possible by the built-in stirrer in the Beckman Acta III spectrophotometer which keeps the enzyme derivative well-suspended throughout the assay period. This method also permits a quick and convenient determination of soluble enzyme in the added immobilized derivative. If the immobilized derivative is free of soluble enzyme turning off the stirrer causes the solid to settle and the cessation of product appearance.

Assay procedure for combined three-enzyme system. Starting with either glyceraldehyde-3-phosphate or dihydroxyacetone phosphate and the three-enzymes, TPI, aldolase, and FDPase, the overall reaction leads to the formation of fructose-6-phosphate and inorganic phosphate. The measurement of reaction rate is based on the appearance of inorganic phosphate.

The assay procedure consists of two phases in which the enzyme solution is first incubated under various conditions and then measured for activity by phosphate determination of an aliquot. For the enzyme incubation solution, either glyceraldehyde-3-phosphate or dihydroxyacetone phosphate could serve as the initial substrate. Dihydroxyacetone phosphate was preferred, however, since all of the added substrate is used as compared to only one-half of the D,L-glyceraldehyde-3-phosphate. The dihydroxyacetone phosphate was obtained commercially as the diethylketal. Before use, the desired amount of the diethylketal was placed in 0.1 N HCl at 40°C for 1 hour. The pH was adjusted to 6.0 and the volume adjusted to give a final concentration of 0.04 to 0.05 M. Actual concentration was verified by assay with glycerophosphate dehydrogenase.

A typical assay solution consisted of the following components: 0.5 ml 0.1 M glycine buffer, pH 9.5, 0.1 ml 4 mM dihydroxyacetone phosphate, 0.1 ml 10 mM $MnCl_2$, 0.3 ml H_2O, and 0.01 ml of each of the three enzymes. At the end of timed intervals, aliquots were removed for phosphate analysis. Phosphate was determined by the Fiske-Subbarow method. The aliquot to be measured was added to a solution of H_2O and 0.2 ml 9.25 N H_2SO_4 to give a volume of 3.2 ml. Samples could be maintained at this step until convenient to develop the color. To develop the color, 0.4 ml 2.5 percent ammonium molybdate was added followed by 0.2 ml reducing agent (made by mixing 0.02 g 1-amino-2-naphthol-4-sulfonic acid, 0.12 g sodium bisulfite, and 0.12 sodium sulfite in 100 ml H_2O). After 10 minutes, the tubes were read at 740 nm. A standard curve was prepared using a series of phosphate standard solutions.

Immobilization. The three enzymes were immobilized separately by the use of gel entrapment, adsorption onto colloidal silica, CNBr-activated Sepharose and glutaraldehyde-treated, alkylamine glass as described earlier. Similar procedures were used to prepare derivatives containing all three enzyme immobilized simultaneously and in varying ratios.

Phosphoramidate-Hexose Transphosphorylase (PHT)

Assay. The assay is based on the fact that phosphoramidate is extremely acid labile and is converted to inorganic phosphate quantitatively by adding a strong acid to stop the reaction. Using phosphoramidate labelled with ^{33}P (prepared according to the procedure of Stokes[10]), the ^{33}Pi after reaction termination, representing unreacted phosphoramidate, was removed by precipitation as the $MgNH_4$ $^{33}PO_4$ salt.

A typical assay mixture consisted of 0.2 ml Tris, 0.5M, pH 8.7; 0.1 ml 0.05 M phosphoramidate and 0.1 ml 0.4 M glucose. The mixture was placed in a 37C water bath for 5 min and the reaction started by adding 0.1 ml enzyme. After 5 min, 0.1 ml 60 percent perchloric acid was added to stop the reaction and hydrolyze unreacted ^{33}P-phosphoramidate. The tube was left in the 37°C bath for an additional 3 min (to complete hydrolysis) and then treated with 0.1 ml 1M magnesium acetate followed by 0.2 ml concentrated NH_4OH. The tube was centrifuged and 0.45 ml supernatant removed for counting. A background blank was made by omitting the enzyme from the reaction mixture. One unit of activity was defined as the quantity of enzyme necessary to convert 1 μmole phosphoramidate to glucose-1-phosphate per minute.

Enzyme purification. The enzyme was isolated from E. coli, Crookes strain (ATCC No. 8739) and purified according to the procedure of Stevens-Clark.[11]

Immobilization. Using enzyme obtained from pooled DEAE-cellulose chromatographic fractions, immobilization was performed using the glutaraldehyde-alkylamine glass technique.

Carbamyl Phosphokinase

Details concerning the isolation, assay and immobilization of this enzyme have been published elsewhere.[12]

Phosphorylase

A description of phosphorylase immobilization has been published elsewhere.[13]

USE OF IMMOBILIZED ENZYMES FOR SYNTHETIC PURPOSES   357

RESULTS AND DISCUSSION

The overall objective of the work described here has been the study of immobilized enzyme systems for the synthesis of polysaccharide from various three-carbon precursors. Initially, the idea was to begin with biochemical CO_2 fixation using ribulose-1,5-diphosphate. Investigation of this pathway in our laboratory started with the two-enzyme and two-cofactor sequence converting 3-phosphoglycerate into glyceraldehyde-3-phosphate. Following a preliminary investigation of this two-enzyme system, an alternate pathway was considered which relied on non-enzymatic reactions for conversion of CO_2 into suitable three-carbon chemicals such as dihydroxyacetone.

Since either of these pathways leads to the formation of triose phosphates, part of our study dealt with the reactions necessary to convert triose phosphate into polysaccharide. This was divided into two segments—conversion of triose phosphate into hexose derivatives and conversion of hexoses into polysaccharide.

Difficulties with the preparation of highly active immobilized glyceraldehyde-3-phosphate dehydrogenase led to the examination of a large number of techniques. Results of the various immobilization procedures for both GAPDH and PGK are summarized in Table 1. With one exception, all of the experiments represented in the table were with crystalline enzymes. With this source, the best activity was obtained with CNBr-activated Sepharose.

TABLE 1. ACTIVITY OF IMMOBILIZED GAPDH AND PGK

		Activity (units by direct assay/g support)	
		GAPDH	PGK
1.	CNBr-activated Sepharose		
	with pure muscle enzyme	16	6
	with impure yeast mixture	100–140	90–110
2.	Porous glass		
	alkylamine glass + glutaraldehyde	4.3	8
	arylamine glass + diazotization	2.5	–
3.	Colloidal silica, crosslinking with glutaraldehyde or carbodiimide	*	–
4.	Adsorption on DEAE & CM Sephadex	*	–
5.	Iodoacetyl cellulose	2.5	–
6.	$TiCl_4$-activated cellulose	0	–
7.	Gel entrapment	0	*

* Activity not stated because the enzyme continually washes off.

Since GAPDH has a number of -SH groups which are essential for activity, some experiments were designed to see if the low activity was due to formation of inactive and irreversible sulfhydryl derivatives. GAPDH was first reacted with 5,5'-dithiobis (2-nitrobenzoic acid) (DTNB) which inactivates the enzyme by forming a mixed disulfide with the essential -SH groups. Exposing the DTNB-treated enzyme to an excess of dithioerythritol reactivates the enzyme by regenerating the free -SH groups. The DTNB-inactivated GAPDH was used as the enzyme source for several immobilization experiments. Following immobilization, the DTNB blocking group was removed and the level of activity checked. This process was found to have little beneficial effect on the activity of the final product. The inactivation-reactivation process itself worked with soluble GAPDH so the reason for poor retention of activity after immobilization must be one of the other possible explanations.

The most striking result in Table 1 is the high activity obtained with a crude yeast enzyme mixture in place of crystalline muscle enzyme. Yeast GAPDH differs from muscle GAPDH in several ways which could make it a better candidate for immobilization. We had crystalline muscle GAPDH but no yeast enzyme. Rather than purchase the crystalline yeast GAPDH and repeat the same immobilization experiments, we prepared partially purified GAPDH from one of the several dried yeast types on hand (Bottom brewer's yeast was far superior to the other types examined such as baker's or Candida utilis). Furthermore, the purification procedure could be adjusted so that high levels of GAPDH and PGK would be present in one solution. The hope was that upon immobilization with such a mixture, both enzymes would be simultaneously attached. As can be seen in Table 1, both enzymes were simultaneously attached and at reasonably high levels. Whether the difference is due to the use of yeast rather than muscle enzymes or the use of a crude mixture cannot be determined from this information. With both enzymes immobilized on the same support, rate enhancements of the type observed by Mosbach[14] were observed. In solution, the activity of the combined enzyme pair is less than the activity of either enzyme measured separately (i.e., in the presence of an excess of one of the pair). However, the immobilized enzyme pair shows the opposite pattern. The activity of the pair measured together is higher than either of the separate assays.

Once we started to use the enzyme pair as a unit, the effect of various parameters such as pH, temperature and substrate concentration had to be investigated. For example, the pH optimum for the operation of the two-enzyme system might be different than that determined for either enzyme alone. We examined buffer type first and found imidazole-HCl to be most effective. It was later determined that chloride ion was important. Imidazole-HCl at a final concentration of 0.05 M was better than 0.02 M. However, addition of NaCl to make chloride 0.05 M restored the activity to the same as 0.05 M imidazole-HCl. The pH optimum for the system was between 6.5-7.0. The activity increased with increasing assay temperature up to 55°C which was the highest temperature used.

The effect of ATP and NADH concentration was studied with the immobilized enzyme pair. In batch-type assay using well-stirred enzyme suspensions, the activity was measured over a range of ATP and NADH concentrations. This type of investigation typically is used to determine the Km for a given enzyme. However, since we are measuring the overall effect of two enzymes, the calculated constant is only an apparent Km. A similar investigation on the effect of ATP concentration was performed with a column of GAPDH/PGK-Sepharose. All of the apparent Km information is summarized in Table 2.

TABLE 2. APPARENT Km FOR ATP AND NADH USING IMMOBILIZED GAPDH/PGK

Substrate		Km (mM)
NADH	Well-stirred suspension	0.088
ATP	Well-stirred suspension	0.37
ATP	Column; 0.2 ml/min	0.27
ATP	Column; 0.3 ml/min	0.22
ATP	Column; 0.4 ml/min	0.18

To study the stability of the immobilized enzyme pair, the most active derivative, PGK/GAPDH-Sepharose, was placed in a small column and operated continuously over several days. The conditions selected represented a compromise between several factors. First, concentrations of each reactant (PGA, ATP, and NADH) were identical so that 100 percent conversions, no excess reactants would remain. This was done in anticipation of product separation processes to be included later. Second, the initial concentration of the reactants was selected to be 0.2 mM so that high 340 nM absorbance (due to NADH) was avoided. This concentration was above the apparent Km for NADH but below the apparent Km for ATP. The apparent Km for PGA in the immobilized system was not determined. An amount of immobilized enzyme was used so that at 0.05-0.5 ml/min flow rates, the response of the column was between 50-100 percent conversion. The important point was to avoid an excess of enzyme for a given concentration of substrates. Using only 0.1 ml of the Sepharose derivative and 0.4 ml untreated Sepharose as a column filler, the response of the column to changes in flow-rate is shown in Table 3.

TABLE 3. PERCENT CONVERSION USING A PGK/GAPDH-SEPHAROSE COLUMN

Flow-rate ml/min	Percent Conversion (Initial concentration of substrates 0.2 mM)
0.1	93.5
0.2	85.0
0.4	69.1
0.5	61.5

This column was then operated at a flow-rate of 0.1 ml/min for several days. From the initial conversion of 93.5 percent, the activity slowly declined as determined from samples collected hourly. After approximately 54 hours continuous operation, the activity of the column had decreased by 50 percent of the initial activity. At the end of 64 hours operation, the substrate solution going through the column was replaced with a solution of soluble PGK/GAPDH (crude yeast preparation). With the soluble enzyme in the column, the flow was stopped and the column stored in the refrigerator overnight. The next day, the soluble enzyme was washed out using 0.05 M imidazole buffer, pH 7.0. Substrate solution was again pumped through the column on a continuous basis. After 22 hours additional operation, the same soluble enzyme treatment was repeated. This time, however, glutaraldehyde was added to the column after washing out excess soluble enzyme but before resuming substrate flow.

The treatment of the column with soluble enzyme solution had a regenerating effect as seen in Figure 3. Before the first soluble enzyme addition, the activity was down to 25 percent of the starting value. After overnight incubation with soluble enzyme, the column activity rose to a level approximately 80 percent of the initial activity. From this point, activity decay was at a rate similar to the latter stages of the column before soluble enzyme treatment. Regeneration was accomplished a second time with a second soluble enzyme addition. At this point, glutaraldehyde was included in an effort to prevent further rapid decay. However, as shown in Figure 3, this treatment had no beneficial effect.

USE OF IMMOBILIZED ENZYMES FOR SYNTHETIC PURPOSES 361

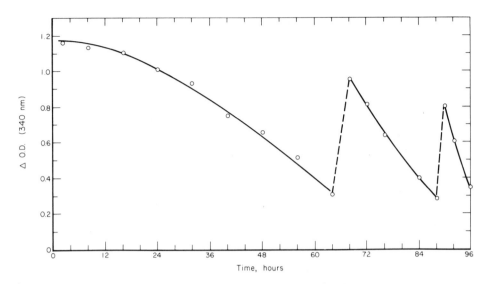

Figure 3. Continuous Operation of a PGK/GAPDH-Sepharose Column

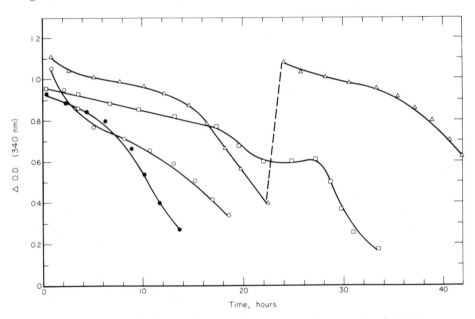

Figure 4. Continuous Operation of Immobilized PGK/GPADH Derivatives
●-● Pure PGK/GAPDH on Sepharose
o-o Crude PGK/GAPDH on Sepharose with glutaraldehyde during coupling
□-□ Crude PGK/GAPDH on glass
△-△ Crude PGK/GAPDH on Sepharose. Glutaraldehyde added after coupling.

The regenerating effect observed with this column system is difficult to explain. This system is complicated by virtue of being a two-enzyme derivative (actually more than two, considering the impure starting material) and also involving subunit enzymes. There is very little previous information concerning the long-term operation of immobilized subunit enzymes. Several possible explanations can be offered, but direct proof of any would be hard to obtain. One possible explanation would be that the two enzymes PGK and GAPDH, are strongly bound either to each other or some other protein in the crude mixture and that covalent attachment to Sepharose occurs through only part of the protein-protein complex. Continuous flow of substrate could then be pictured as slowly removing the noncovalently bound part of any such protein-protein complexes. Adding back the soluble enzyme solution would permit restoration of the active complex. Amino acid analysis of the immobilized derivative before and after long-term use might reveal protein losses but this was not performed. Arguing against this explanation is the result obtained with pure PGK/GAPDH in which no other proteins were present. As seen in Figure 4, immobilized PGK/GAPDH obtained from pure yeast enzymes decayed at a rate even faster than the preparation made with a crude mixture. A second explanation is that attachment to the support is through only one of the enzyme's subunits and that subsequent subunit dissociation leads to a loss in activity. Working with immobilized aldolase, a tetrameric subunit enzyme, Chan and Mawer[15] showed that treatment with reagents favoring subunit dissociation led to loss of activity which could be partially recovered when exposed to soluble aldolase. In our experiment with PGK/GAPDH-Sepharose (Figure 4), glutaraldehyde was added following column regeneration in an effort to crosslink subunits and prevent subsequent loss of activity. This treatment, however, did not prevent the rapid decay of activity.

To further examine the stability of immobilized PGK/GAPDH, different types of derivatives were prepared. Results of continuous column operation with these preparations are shown in Figure 4. There were variations in the stability of the different preparations but none of them showed greater stability than the original Sepharose column.

One additional effort to immobilize PGK/GAPDH involved the microencapsulation technique. With the enzyme retained inside a semi-permeable membrane, escape of protein during continuous use would be prevented. However, the activity of the enzyme pair surviving the encapsulation process was too little to be of value. We encapsulated the $(NH_4)_2SO_4$ precipitate as a suspension in 70 percent saturated $(NH_4)_2SO_4$. To gradually remove the $(NH_4)_2SO_4$, the capsules were placed in a 70 percent saturated $(NH_4)_2SO_4$ solution which was slowly diluted with H_2O.

At 5 percent saturation, the capsules were filtered, washed with H_2O and resuspended in 0.05 M imidazole buffer, pH 7.0. An assay was conducted by the batch and column process. The activity was approximately 0.1 unit/g capsules. One reason for the observed low activity may be due to limited permeability of the substrates across the membrane. Supporting this idea is the observation that capsule rupture leads to a release of active enzyme in an amount greater than expected from assay of the intact capsules.

The next phase of this investigation was concerned with the enzymes involved in the conversion of triose phosphate into fructose-6-phosphate. The three enzymes - triose phosphate isomerase, aldolase, and fructose diphosphatase - were immobilized both separately and together by a variety of techniques. The results of a large number of immobilization experiments are summarized in Table 4.

The optimum ratio of TPI:aldolase:FDPase was determined by an empirical approach in which various amounts of each enzyme were mixed together and the overall rate of reaction determined (i.e., rate of appearance of phosphate). Results of a series of experiments in which the TPI:aldolase:FDPase ratio was varied are shown in Table 5.

An excess of FDPase relative to aldolase seems necessary for optimum activity. From the information in Table 5, a desired ratio of the three enzymes is approximately 5:1:2. The ratio, while not the optimum, permits reasonably rapid synthesis with a minimum of each enzyme. The information from this experiment was considered to be a useful beginning guide for the preparation of various immobilized three-enzyme derivatives.

We then attempted to immobilize the three enzymes simultaneously hoping to arrive at a final activity in the approximate ratio of 5:1:2. As seen in Table 7, the actual ratio obtained with Sepharose was 5:4:1. This contains too much aldolase and too little FDPase. The initial specific activity of the aldolase was 10 units/mg compared to only 1 unit/mg for FDPase. This means that to get a higher activity of FDPase compared to aldolase requires considerably more FDPase protein than aldolase protein. Loading the Sepharose with the desired amount of FDPase, however, reduces the capacity of Sepharose to bind TPI and aldolase. One solution to the problem would be to find supports which bind the three in a more favorable ratio. Another solution would be to begin with more active FDPase thus requiring smaller amounts of protein.

Efforts toward finding a better support for the three-enzyme mixture included several types of covalent attachment, physical adsorption, and gel entrapment. With both alkylamine and arylamine

TABLE 4. ACTIVITY OF IMMOBILIZED TPI, ALDOLASE, AND FDPase

Enzyme	Immobilized Separately	
	Method	Activity, units/g dry solid
TPI	Alkylamine glass	61
	Arlamine glass	(a)
	Sepharose	8-30
	Gel entrapment	1
Aldolase	Alkylamine glass	4.5-6
	Arlamine glass	3.5
	Sepharose	29
	Silica	(b)
	Gel entrapment	0.3
FDPase	Alkylamine glass	8
	Arylamine glass	8
	Sepharose	6
	Silica	(b)
	Gel entrapment	0.8

	Immobilized Together		
	Enzyme Activity, units/g dry solid		
Method	TPI	Aldolase	FDPase
Alkylamine glass	(a)	18	4
Gel entrapment	0.5	0.25	0.75
Sepharose	30	22	6

(a) Enzyme continually washing off.
(b) Binds well at pH 7 but comes off at assay pH (9.4).

TABLE 5. ACTIVITY OF VARIOUS COMBINATIONS OF
TPI, ALDOLASE, AND FDPase

Enzyme Ratio TPI:Aldolase:FDPase	Activity µmoles phosphate/5 min
10:1:10	0.752
10:1:8	0.648
10:1:6	0.635
10:1:4	0.595
20:1:2	0.579
10:1:2	0.566
6:1:2	0.524
4:1:2	0.491
10:1:1	0.358
10 10:1	0.358
1:1:10	0.279
1:10:10	0.233
1:10:1	0.177

glass, considerable difficulty was encountered trying to wash the glass free of noncovalently bound TPI. With colloidal silica, good binding was obtained in pH 7-7.5 range but the enzymes would come off at pH 9.4 despite crosslinking with glutaraldehyde. Gel entrapment led to preparations with low activity. CNBr-activated Sepharose, therefore, seems to be the best of the methods examined.

The other solution mentioned, i.e., obtaining more active FDPase, is a distinct possibility, but remains to be investigated. The current source of FDPase is of commercial orgin and in the range of 1-2 units/mg specific activity. Some recent work[16] indicates that a rapid purification of the enzyme is possible from rabbit liver giving specific activities in the range of 5-10 units/mg. Furthermore, research from the group of Horecker, Pontremoli, and coworkers[17] has shown that the specific activity can be increased to near 40 units/mg by treatment with certain disulfides such as homocystine. A specific activity of 30-40 units/mg should permit the preparation of a Sepharose derivative with an activity ratio more nearly approaching the desired range.

The last step in either of the major pathways is the phosphorylase-catalyzed formation of polysaccharide. Although the last step, we actually investigated this reaction before the work just described. This work has been published[13] and need not be repeated in detail here. The major finding was that phosphorylase could be immobilized in a straightforward manner with good retention of activity. A column of phosphorylase attached to alkylamine glass was operated continuously for nearly a month. Polysaccharide of varying molecular weight was obtained by varying the glucose-1-phosphate : primer ratio in the substrate feed solution.

The study of the enzyme phosphoramidate-hexose transphosphorylase (PHT) was of interest since it produces glucose-1-phosphate directly. Previous studies[11] have shown that several reactions can be catalyzed by the same enzyme as shown below.

glucose + phosphoramidate ⟶ glucose-1-P + NH_3

glucose + phosphoramidate ⟶ glucose-6-P + NH_3

phosphoramidate + H_2O ⟶ Pi + NH_3

glucose-1-phosphate ⟶ glucose + Pi

With purified soluble enzyme preparations, incubating phosphoramidate and glucose led to the formation of glucose-1-P which reached a maximum and then began to decline. Presumably this reflects the phosphatase activity of the preparation. The immobilized enzyme preparation (60-70 units/g glass) used in column form gave similar results. It was found that decreasing the flow rate and thereby increasing the contact time caused an increase in the percent conversion into glucose-1-P up to a maximum value (50-60 percent conversion). Further decreases in flow rate led to a smaller production of glucose-1-P. Of the different reactions catalyzed by PHT, the two which appear to offer the most serious competition to the production of glucose-1-P are the hydrolysis of phosphoramidate and the hydrolysis of glucose-1-P. The latter reaction could be demonstrated by passing a glucose-1-P solution without phosphoramidate through the enzyme column and determining the amount of inorganic phosphate in the effluent. The phosphatase activity was quite severe at pH 7.5 but was reduced considerably at pH 8.5 (from 93 percent hydrolysis to 37 percent at 0.2 ml/min flow rate and from 50 percent down to 20 percent at 0.8 ml/min). No effort was made to measure only the phosphoramidase activity of the preparation. This competing reaction, however, is probably the most serious side-reaction, especially under conditions which minimize glucose-1-phosphatase activity.

To maximize the yield of glucose-1-P, parameters such as substrate ratio, pH, temperature and flow rate were examined. The maximum initial reaction rate was obtained at pH 7.5, but a pH of 8.5-9.0 was preferred to minimize phosphatase activity. Temperature optimum for the immobilized derivative was 60°C at which point the soluble derivative was only 40 percent as active. The optimum substrate ratio was 1:4 (phosphoramidate : glucose).

The stability of the immobilized PHT derivative was tested by continuous operation. With 0.25 M glucose, 0.6 M phosphoramidate, 0.14 ml/min, 45°C and pH 9.0, the activity declined rapidly. After two days continuous operation, the percent conversion was approximately one-half the starting value. Solubility of the glass support at this pH could account for much of the rapid decline. A similar experiment at pH 7.5 showed an observed half-life of approximately 7 days.

A necessary part of any of the pathways considered was a means of ATP regeneration. We considered several possible methods and finally selected the reaction catalyzed by carbamyl phosphokinase.

The phosphoryl donor, carbamyl phosphate, is easily made by aqueous reaction between cyanate and phosphate. Furthermore, the reaction product, carbamate can decompose to gaseous products which could aid in the separation and recovery of the desired product, ATP.

A preliminary study of this regeneration scheme has been published.[10] Although it does show the feasibility of the proposed method, much additional study is needed to determine the suitability of this method over some of the other alternatives (such as the acetyl phosphate route[18]).

Acknowledgement

The author would like to thank Mrs. Melody K. Bean for excellent technical assistance and Dr. Richard Falb for advice and encouragement during this investigation. This work was supported by the National Aeronautics and Space Administration, Contract No. NAS-2-5956.

References

1. J. Porath, R. Axen, and S. Ernbach, Nature, 215, 1491 (1967).

2. R. Haynes and K. Walsh, Biochem. Biophys. Res. Commun., 36, 235 (1969).

3. P. Robinson, P. Dunhill, and M. Lilly, Biochem. Biophys. Acta, 242 659 (1971).

4. T. Sato, T. Mori, T. Tosa, and I. Chibata, Arch. Biochem. Biophys., 147, 788 (1971).

5. A. Patchornik, U.S. Patent 3,278,392 (1966).

6. A. N. Emery, The Chem. Engineer, 71 (1972).

7. H. Nilsson, R. Mosbach, and K. Mosbach, Biochem. Biophys. Acta, 268, 253 (1972).

8. J. K. Inman and H. M. Dintzis, Biochemistry, 8, 4074 (1969).

9. D. L. Gardner, R. D. Falb, B. C. Kim, and D. C. Emmerling, Trans. Am. Soc. Artif. Organs, 17, 239 (1971).

10. H. N. Stokes, Am. Chem. J., 15, 198 (1973).

11. J. R. Stevens-Clark, M. C. Theisen, K. A. Conklin, and R. A. Smith, J. Biol. Chem., 243, 4468 (1968).

12. D. L. Marshall, Biotech. Bioeng., 15, 447 (1973).

13. D. L. Marshall and J. L. Walter, Carbohyd. Res., 25, 489 (1972).

14. K. Mosbach and B. Mattiasson, Acta Chem. Scand., 24, 2093 (1970).

15. W. W. C. Chan and H. M. Mawer, Arch. Biochem. Biophys., 149, 136 (1972).

16. S. Traniello, E. Melloni, S. Pontremoli, C. Sia, and B. Horecker, Arch. Biochem. Biophys., 149, 222 (1972).

17. K. Nakashima, B. Horecker, and S. Pontremoli, Arch. Biochem. Biophys., 141, 579 (1970).

18. C. K. Colton and G. M. Whitesides, 40th Annual Chemical Engineering Symposium, Purdue University, Jan. 23-24, 1974.

INDEX

A

Acetoacetatesuccinyl CoA transferase, binding of Blue Dextran 2000 123
Acetylcholinesterase, purification of 75
Activity of immobilized enzymes,
  effect of temperature 204ff
  effect of time 202ff
Adenosine 5'-monophosphate 85
Adenosine 5'-triphosphate 85
Affi-Gel 10, as affinity matrix 169
Affinity chromatography, acetylcholinesterase 75
  aldehyde oxidase 165
  allosteric effectors 85
  analysis of particulate systems 4
  antibody-antigen interaction 5
  dehydrogenases 99ff
  effect of temperature 111
  fructose 1,6-diphosphatase 85
  GTP-ring-opening enzyme 147
  labeled peptides 23
  lymphocyte antibodies 5
  membrane constituents 5, 6
  ornithine transcarbamylase 61
  ovalbumin 20
  probes of cell surfaces 5, 6
  protein kinases 17, 19
  purification of dehydrogenases 9
  purification of estradiol acceptors 6
  purification of FSH and LH 5
  purification of β-galactosidase 4
  purification of haptens 24
  purification of heavy meromysin 22
  purification of polyribosomes 5

purification of rhodopsin 5
purification of viruses 5
purification of vitamin B_{12} binding protein 6
pyridoxal phosphate as a ligand 5
quantitative parameters 33ff
radioimmune assay 61
separation of peptides 27
thymidylate synthetase 135
trypsin 36
tyrosyl-tRNA synthetase 157
use in protein sequencing 27
xanthine oxidase 165

Affinity labeled peptides, purification of 23

Affinity ligands 15ff
antibody 3
binding to 34
concanavalin A 5
cyclic AMP 16, 17, 19
dyes 123
5-fluoro-2'-deoxyuridine 5'-phosphates 135
hormone 3
inhibitor 3
insulin 9
isoproterenol 6
leakage 8, 19
NAD^+ 9
neutiminidase inhibitor 5
organophosphonates 76
partially hydrolyzed Vitamin B_{12} 6
plant lectins 5
pyridine nucleotides 99
pyridoxal phosphate 5
17-substituted estradiol 6
subunit 3
N^6 succinyl cyclic AMP 19
use of macromolecular spacer arms 10
use of spacers 9, 10, 15ff

Agar, as a matrix for nucleic acid entrapment 177

Agarose, as a matrix for immobilizing lipase 340

Agarose, carboxyl linked tyrosine 157
charge-free derivatives 15ff
coupling of alanine 8
coupling of albumin 8
coupling of hydrazides 20
coupling of RNA and UMP 147
cyanogen bromide activation 4, 7, 8, 10, 11, 18
DNA derivatives 181
insulin, interaction with fat cells 4, 5
preparation of hydrazide-albumin derivatives 10
use of albumin and poly-L-lysine-alanine as spacers 6

Albumin, binding to Blue Sepharose 128
bovine serum 44
human serum 29
receptor 5
use as a spacer 10

Alcohol dehydrogenase 100, 103, 106
binding to Blue Sepharose 131

Aldehyde oxidase, affinity chromatography 165

Aldolase, binding to Blue Sepharose 128
immobilization on various matrices 354-356

Allosteric effectors 85

Alkaline phosphatase, analytical uses 207
half-life when immobilized 203
immobilized on glass 200

Alkyl nucleotides, as affinity ligands 99

Alkyl Sepharoses 20

INDEX

Alkylamine-glass 191
Alkylamine Sepharose 43ff
Alumina, as a matrix for enzyme immobilization 214
Amethopterin 135
L-Amino acid oxidase, half-life when immobilized 203
  immobilized on glass 200
Aminoacyl t-RNA synthetases 181
Aminoalkyl-AMP analogues 100
Aminoalkyl-Sepharose 16
6-Aminocaproic acid, as a spacer arm 36
AMP, immobilized derivatives 187
N^6 AMP-Sepharose 100
Amylase 3
Amylases, immobilized on Enzaacryls 293
Amyloglucosidase, immobilization 188
Analytical ultracentrifugation 81
Antibody-antigen interactions 5, 61, 66
Antibody-Sepharose column 24, 26
Anti-DNP antibodies 24ff
Anti-DNP-Sepharose column 24, 26
Arginine biosynthetic regulon 61
Arrenhius plot, immobilized lactase 197
Arylsulfatase, immobilized on glass 200
L-Asparaginase, immobilized on polymethyacrylate 208
Asparaginase, immobilization on stainless steel 221
ATP regeneration, immobilized enzymes 348
ATP Sepharose hydrazide 22
Automated peptide synthesizer 286
Azo coupling of proteins 192

B

Bacitracin, binding to Blue Sepharose 128
Bacteriophage T4DNA, coupling to phosphocellulocse 179
Batch reactors 201
Bed reactors 213
N-Benzyl nicotinamide, inhibition of aldehyde oxidase 165
Bio-gel P-150 hydrazide 351
Blood coagulation factors, binding of Blue Dextran 2000 123
Blue Dextran 2000, binding to proteins 123
Blue Sephadex 124
Blue Sepharose 124

C

Carbamyl phosphokinase, immobilization of 356
Carbodiimide coupling 10, 36, 137
  glass 192
  polynucleotides 180
  thymidine 5'-phosphate 174
Catalase, immobilization on NiO and alumina 221
Cellulose 3
  acetate, as a matrix for immobilization 177
  aminoethyl 175
  immobilization of nucleic acids 176
  iodoacetyl 350
  matrix for immobilizing lipase 339
  phosphate 85
  TiCl$_4$-activated 351
Chymotrypsin, hydrophobic chromatography 43, 58
  immobilized 293, 329
  immobilized on Sepharose 269
Chymotrypsinogen, binding to Blue Sepharose 128
  immobilized on glass beads 333
Circular dichroism spectra 298

Citrate synthase, immobilization 188
Cofactors, immobilized 187
Collagen, as a matrix for enzyme immobilization 319
Concanavalin A 5
Coupling agents, silanes 191
Coupling, ligands 7, 136
 with glutaraldehyde 6
Covalent affinity chromatography 75
Creatine kinase, binding to Blue Sepharose 125
Cyanoethyl phosphate 137
Cyclic AMP 16
 as an affinity ligand 187
Cytochrome C 29
 binding to Blue Sepharose 128

D

Dehydrogenases, affinity chromatography of 123
Denaturation - renaturation of immobilized enzymes 331ff
3-Deoxy-D-arabinoheptulosonate 7-phosphate synthetase, purification of 157, 160
Deoxynucleotidyl transferase 180
Deoxyribonuclease 180
DNA agarose 181
 cellulose 181
 immobilization of single stranded 176
 polymerase 180, 181
Diazotization, coupling 193
Dihydrofolate reductase 135
Dihydroxyboryl groups, polynucleotide immobilization 175
Diisopropylphosphofluoridate 336

E

Elution of affinity columns 9
Encapsulation of enzymes 352
Endonuclease 180, 181
Enzyme entrapment in polyacrylamide gels 351
Enzyme immobilization, criteria for 242
 loss of activity 247
 on inorganic supports 191
Enzyme ligand binding 34ff
 modification reactions 309
 pH electrodes 188
 technology 345
 water encapsulated 259
Equilibrium binding studies 36, 37
Estradiol receptor purification use of spacers 10
Estrogen receptor protein 8
Exonuclease I and II 181

F

Ficin, immobilized on CM-cellulose 269
Fixed bed reactors 201
Fluidized beds 214
Fluorescence emission spectra, immobilized enzyme 337
5-Fluorodeoxyuridine 3',5'-diphosphate, derivatives 137ff
5-Fluoro-2'-deoxyuridine 5'-phosphate, as an affinity ligand 135
5-Fluorouracil 135
Fructose 1,6-diphosphatase, elution studies 85
 immobilization on various matrices 354-356

G

β-Galactosidase 4, 235
 immobilization 188
Gel filtration, dissociation constants 92
Gene isolation 179
Glass, aminoalkyl 191, 284, 330, 350
 beads, matrix for peptide synthesis 283
 coating with zirconium salts 249
 controlled pore 241
 matrix 7
 matrix for lactate dehydro-

genase 330
reaction with silanes 191
ZrO$_2$-coated 195
γ-Globulin 44
Glucoamylase, half-life when
immobilized 202, 241
immobilized on glass 200
pH optimum of immobilized
enzyme 248
Gluconeogenesis 94
Glucose 6-phosphate dehydrogenase 100, 103, 107
adsorption 22
binding to Blue Sepharose
131
immobilization 188
Glucose oxidase, immobilization
188
immobilized on glass 200
Glutamate dehydrogenase 108,
109
Glutaraldehyde, immobilization
of proteins 6, 193
coupling of enzymes to
metals 220
Glutathione reductase 100
binding of Blue Dextran
2000 123
Glyceraldehyde 3-phosphate
dehydrogenase 103
binding to Blue Sepharose
131
immobilized on Sepharose
348
immobilized on silica 349
Glycerokinase 103, 106
Glycinamide, reaction with matrix bound carboxyl groups
144
GTP, coupling to Sepharose 148
GTP ring opening enzyme, affinity chromatography 147
properties 153

H
Half-lives, immobilized enzymes
202
Hemocyanin receptor 5

Hemoglobulin, binding to Blue
Sepharose 128
Hexokinase 103
binding to Blue Sepharose
131
immobilization 188
Hormone-receptor systems 8
Hydrazido Sepharose 20, 175
Hydrophobic columns 7
chromatography 21, 43ff
spacers 43ff
N-Hydroxysuccinimide, as a
spacer 10, 36

I
Immobilization, polynucleotides
173
Immobilized enzyme, analytical
uses 206, 207
ATP regeneration 348
chain refolding 329
durability tests 251
economic potential 206
optimization of activity
293
reaction rates 270
reactor 220, 235
subunit interactions 329
synthetic uses 345
systems 241
therapeutic uses 208
Insulin receptor 5
Invertase, immobilized on glass
200
Iodoacetylcellulose 350
Isocitrate dehydrogenase 108

K
Kinases affinity chromatography
of 123

L
Lactases, half-lives when immobilized 202
Lactase, immobilized on glass
194
immobilized on metals 213
immobilized on stainless
steel and alumina 221

membrane system   235
Lactate dehydrogenase   103, 104ff
   binding to Blue Sepharose   130, 131
   immobilization   188, 329
   isozymes   187
β-Lactoglobulin   44
Langmuir adsorption isotherm   34
Lectins   5
Lipase, immobilized   293
   immobilized on polysaccharide matrices   339
Liquid distributor, for reactors   229
Lymphocytes, fluorescence antibody   9
Lysozyme   29
   immobilized on polysaccharide   294

M

Malate dehydrogenase   100
   immobilization   188
Mass transfer coefficient   218
Membrane bound enzymes   235
   ultrafiltration   261
Metal oxides, enzymes immobilized on   213
   matrix   191
Methylene tetrahydrofolate   135
N-Methyl nicotinamide, substrate for aldehyde oxidase   165
N-Methylpyrrolidone, as a coupling agent   36
Michaelis-Menten kinetics   40, 271
Microenvironments, solid state   283
Multi-step enzyme systems   187
Murine mammary cells   9

N

NAD^+, immobilized derivatives   187
   Sepharose columns   101
NADH, eluant for lactate dehydrogenase   113
   Km with immobilized lactate dehydrogenase   330
Nickel oxide, as a matrix for enzyme immobilization   214
Nuclease-Sepharose columns   23
Nucleic acid enzymes, fractionation of   177
   immobilization, UV-irradiation   176
   immobilized   173
   purification of   173
Nylon, as a matrix for nucleic acid immobilization   176
   fibers, as a matrix for antigen   5

O

Oligo (dT)-cellulose   178
   template for RNA polymerase   180
Oligoribonucleotides, fractionation of   178
Organophosphonates   75
Ornithine transcarbamylase, affinity chromatography   61
   purification   64, 65
Ovalbumin, binding to Blue Sepharose   128

P

Papain   29
   half-life when immobilized   203
   immobilized on glass   192
Partition coefficients in affinity chromatography   33
Pepsin, immobilized on glass   192
Peptide synthesis, solid phase   283
Periodate oxidation   149, 175
Phage T4 DNA, immobilization   176
Phosphocellulose, coupling of T4 DNA   179
   matrix for nucleotide immobilization   174
Phosphodiesterase, snake venom   138

Phosphofructokinase, binding of Blue Dextran 2000 123
3-Phosphoglycerate kinase 103
Phosphoglycerate kinase, immobilization 348
Phosphoramidate-hexose transphosphorylase, immobilization on alkylamine glass 356
Phosphorylase, immobilization of 356
  polysaccharide synthesis 348
Phosphorylation of glucose 348
Polyacrylamide, as a matrix for lysozyme 294
  as a matrix for wheat germ agglutinin 294
  Blue Dextran immobilized in 123
  entrapment of nucleic acids 177
  gel electrophoresis 143
  t-RNA complexes 179
Polyacrylamine 7
Polyacrylic hydrazide-Sepharose 22
Poly-DL-alanyl-polylysine as a spacer 18, 19
Poly I-agarose 180
Polylysine, as a spacer 18, 19
  alanine spacers 6, 10
Polymerization trapping 7
Polynucleotides, fractionation 177
  immobilized 173
  kinase 180
  ligase 180
Polyribosomes 5
Polysaccharide, as a matrix for polynucleotide immobilization 173
  enzyme derivatives 269
  synthesis, use of immobilized phosphorylase 348
Poly U-cellulose 179
Polyvinyl beads, immobilization of nucleic acids 176

Polyvinyl binds, coupled poly U 179
Ponceau S 44
Pore diffusion 196
Porous materials, particle size 241
  Pore morphology 241
  surface area 241, 246
Porous support, immobilized enzymes 241
Protamine kinase, affinity chromatography 187
Proteins, azo method of immobilization 193
  carbodiimide method of immobilization 192
  glutaraldehyde method of immobilization 193
  nonspecific binding of 43
  Sepharose columns 23
Proton magnetic resonance, studies of phosphorylated 5-fluorodeoxyuridines 139
Pullanase, immobilized on acrylic copolymer 294
Pyridine nucleotides, as affinity ligands 99
  immobilized 187
  Sepharose 101
Pyridoxal phosphate 5
  with fructose 1,6-diphosphatase 94
Pyruvate kinase 103
  binding of Blue Dextran 2000 123, 130
  binding to Blue Sepharose 125

R

Radioimmune assay, solid phase 61
Reactor, continuous flow 263
  oil-continuous 259
Rhodopsin 5
Riboflavin, biosynthesis 147
Ribonuclease, binding to Blue Sepharose 129
  denaturation 329

immobilized on Sepharose 269
Sepharose column 24
Ribosomal RNA-agarose 179
immobilization 177
RNA, coupling to agarose 147
immobilization of single stranded 176
m-RNA, isolation 178
t-RNA immobilization 175
polyacrylamide complex 179
t-RNA (ser), purification 179
RNA from SV40, immobilization of 178
RNA polymerse 180, 181

S

Saturation capacities 38
Sephadex, as a matrix affinity chromatography 123
as a matrix for immobilized lipase 339
coupled with Cibacron blue 123
Sepharose, alkyl nucleotide derivatives 100, 101
alkylated 20
amino alkyl derivatives 16, 43, 49
AMP derivatives 187
antibody derivative 26
anti-DNP derivative 24ff
CNBr activated 10
CNBr activated, reaction with N-methylpyrrolidone 36
complex amino spacers 136
cyclic AMP derivative 17, 19
derivatives, ion exchange properties 21, 142
dye derivatives 123
5-fluoro-2'-deoxyuridine 5'-phosphate derivatives 135
GTP 148
hydrazide, $NADP^+$ as a ligand 22
hydrazide 20

ornithine transcarbamylase derivative 66
poly-DL-alanyl-polylysine derivative 18, 19
polylysine as a spacer 18, 19
procedure for coupling nucleotides 148
pyridine nucleotide derivatives 101, 187
ribonuclease derivative 24
staphylococcal nuclease derivative 24
substituted nicotinamide derivatives 166
succinyl-aminohexyl 139
Seryl t-RNA, purification 179
Silane, aminoalkyl 284
Solid phase peptide synthesis 283
Spacer groups, albumin 10
6-aminocaproic acid 36
complex amines 136
diamines 16
hydrophobic 43ff
N-hydroxysuccinimide 10, 36
4,4'-methylene dianiline 343
polylysinealanine 6, 10
relation of length to purification 169
succinic anhydride 192
Starch, insoluble 3
Steel, as a matrix for enzyme immobilization 213
Succinyl-aminohexyl-Sepharose 139
N^6-Succinyl cyclic AMP, as a ligand 19

T

Temperature, effects on immobilized lactase 222
n-Tetradecyl alcohol, oxidation 259
Thymidine 3',5'-diphosphate, use in affinity labeling 24

Thymidine dodecanucleotide 178
Thymidylate synthetase, affinity chromatography 135
Thyroglobulin, binding to Blue Sepharose 128
Triose phosphate isomerase, immobilization on various matrices 354-356
Trypsin affinity system 36
Trypsin, immobilization 188
  immobilization on NiO and alumina 221
Tyrosinase 3
  chemical modification 317
  plug-flow reactor 319
Tyrosine aminotransferase 5
Tyrosine-N-carboxyanhydride 158
Tyrosine Sepharose 157
Tyrosyl-t-RNA synthetase, affinity chromatography 157

U

UMP, coupling to agarose 147
Urease, immobilized on glass 200
Uricase, immobilized on Sepharose 269

V

Viral RNA, immobilization 177
Viruses 5

W

Wheat germ agglutinin, coupled to polyacrylamide 294

X

Xanthine oxidase, affinity chromatography 165

Z

Zymogen, immobilized 336

QP519.7 .S9 1973
c.2
Symposium on Affinity
Chromatography and
Immobilized Biochemicals,
Charleston, S.C., 1973
Immobilized biochemicals and
affinity chromatography